NUREG-1887

RASCAL 3.0.5: Description of Models and Methods

Manuscript Completed: August 2007
Date Published: August 2007

Prepared by
S.A. McGuire[a]
J.V. Ramsdell, Jr.[b]
G.F. Athey[c]

[a]Office of Nuclear Security and Incident Response
U.S. Nuclear Regulatory Commission
Washington, DC 20555-0001

[b]Pacific Northwest National Laboratory
P.O. Box 999
Richland, WA 99352

[c]Athey Consulting
P.O. Box 178
Charles Town WV 25414-0178

Prepared for
Office of Nuclear Security and Incident Response
U.S. Nuclear Regulatory Commission
Washington, DC 20555-0001

AVAILABILITY OF REFERENCE MATERIALS
IN NRC PUBLICATIONS

NRC Reference Material

As of November 1999, you may electronically access NUREG-series publications and other NRC records at NRC's Public Electronic Reading Room at http://www.nrc.gov/reading-rm.html.
Publicly released records include, to name a few, NUREG-series publications; *Federal Register* notices; applicant, licensee, and vendor documents and correspondence; NRC correspondence and internal memoranda; bulletins and information notices; inspection and investigative reports; licensee event reports; and Commission papers and their attachments.

NRC publications in the NUREG series, NRC regulations, and *Title 10, Energy*, in the Code of *Federal Regulations* may also be purchased from one of these two sources.
1. The Superintendent of Documents
   U.S. Government Printing Office
   Mail Stop SSOP
   Washington, DC 20402–0001
   Internet: bookstore.gpo.gov
   Telephone: 202-512-1800
   Fax: 202-512-2250
2. The National Technical Information Service
   Springfield, VA 22161–0002
   www.ntis.gov
   1–800–553–6847 or, locally, 703–605–6000

A single copy of each NRC draft report for comment is available free, to the extent of supply, upon written request as follows:
Address: U.S. Nuclear Regulatory Commission
           Office of Administration
           Mail, Distribution and Messenger Team
           Washington, DC 20555-0001
E-mail: DISTRIBUTION@nrc.gov
Facsimile: 301–415–2289

Some publications in the NUREG series that are posted at NRC's Web site address http://www.nrc.gov/reading-rm/doc-collections/nuregs are updated periodically and may differ from the last printed version. Although references to material found on a Web site bear the date the material was accessed, the material available on the date cited may subsequently be removed from the site.

Non-NRC Reference Material

Documents available from public and special technical libraries include all open literature items, such as books, journal articles, and transactions, *Federal Register* notices, Federal and State legislation, and congressional reports. Such documents as theses, dissertations, foreign reports and translations, and non-NRC conference proceedings may be purchased from their sponsoring organization.

Copies of industry codes and standards used in a substantive manner in the NRC regulatory process are maintained at—
    The NRC Technical Library
    Two White Flint North
    11545 Rockville Pike
    Rockville, MD 20852–2738

These standards are available in the library for reference use by the public. Codes and standards are usually copyrighted and may be purchased from the originating organization or, if they are American National Standards, from—
    American National Standards Institute
    11 West 42nd Street
    New York, NY 10036–8002
    www.ansi.org
    212–642–4900

Legally binding regulatory requirements are stated only in laws; NRC regulations; licenses, including technical specifications; or orders, not in NUREG-series publications. The views expressed in contractor-prepared publications in this series are not necessarily those of the NRC.

The NUREG series comprises (1) technical and administrative reports and books prepared by the staff (NUREG–XXXX) or agency contractors (NUREG/CR–XXXX), (2) proceedings of conferences (NUREG/CP–XXXX), (3) reports resulting from international agreements (NUREG/IA–XXXX), (4) brochures (NUREG/BR–XXXX), and (5) compilations of legal decisions and orders of the Commission and Atomic and Safety Licensing Boards and of Directors' decisions under Section 2.206 of NRC's regulations (NUREG–0750).

Abstract

The code currently used by NRC's emergency operations center for making dose projections for atmospheric releases during radiological emergencies is RASCAL version 3.0.5 (Radiological Assessment System for Consequence AnaLysis). This code was developed by NRC. The first version was created about 20 years ago. Since then the code has been undergoing continual improvement to expand its capabilities and to update the models used in its calculations. This report describes the models and calculational methods used in RASCAL 3.0.5. This report updates and supercedes the information in NUREG-1741, "RASCAL 3.0: Description of Models and Methods," 2001.

RASCAL 3.0.5 evaluates releases from: nuclear power plants, spent fuel storage pools and casks, fuel cycle facilities, and radioactive material handling facilities.

While RASCAL 3.0.5 operates as a single piece of software, it is really a set of inter-linked modules each with a different function. These are:

1. Source term: this module calculates a time-dependent source term, which for nuclear power plants, is composed of about 50 radionuclides including parents and daughters. This module is unique in the world for its ability to model a wide variety of accidents based on plant conditions for many different facility types.
2. Meteorological data processor: this module inputs weather observations and forecasts along with local topography to generate time-dependent wind fields that will transport the plume.
3. Atmospheric transport and diffusion: this module uses the wind fields with a two-dimensional Gaussian puff model to transport the plume downwind and to calculate concentrations of each radionuclide as a function of time and location.
4. Dose calculator: this module calculates various types of doses resulting from airborne releases (TEDE, thyroid, acute, etc.) to individuals at each location from three dose pathways - inhalation, cloudshine, and groundshine. It also calculates the longer-term intermediate phase doses from deposited radionuclides. The calculations are completely consistent with the EPA protective action guide manual and the methods adopted by the Federal Radiological Monitoring and Assessment Center (FRMAC).
5. Display of results: this module allows the user to display a wide variety of calculated results as either a picture of the plume footprint on a map background for each of the result types or as numeric table.
6. Uranium hexafluoride module: for uranium hexafluoride releases, RASCAL contains a heavy gas model to account for the exothermic reaction with air and gravitational slumping of the plume.

Table of Contents

List of Figures

List of Tables

Acknowledgments

The development of the RASCAL code would not have been possible without the dedicated work of many talented individuals. Thomas McKenna, Joseph Giitter, and Len Soffer played an important role in developing the models used in the early versions of RASCAL. Frank Congel, Joseph Holonich, Edward Jordan, Melvyn Leach, Charles Miller, Aby Mohseni, Cheryl Trottier, Mike Weber, Richard Wessman, Peter Wilson, and Roy Zimmerman all enthusiastically supported the project.

Andrea Sjoreen and Christian Fosmire contributed to the development of earlier versions of RASCAL.

A large number of people have contributed to the more recent development of RASCAL during its transition from RASCAL 3.0 introduced in 2001 to RASCAL 3.0.5 introduced in December 2006. They include: Bob Bores, Lou Brandon, Mark Cunningham, Paul Elkman, Michelle Hart, Rick Hasselberg, Felicia Hinson, Tony Huffert, Cyndi Jones, Elaine Keegan, Steve Klementowicz, Terry Kraus, Steve LaVie, Bob Meck, Jocelyn Mitchell, Carlos Navarro, Ron Nimitz, Nancy Osgood, John Parillo, Bill Rhodes, Jason Schaperow, Art Shanks, Fritz Sturz, Randy Sullivan, Elizabeth Thompson, Charles Tinkler, Bruce Watson, and Michael Williamson.

We would also like to thank the many RASCAL users and students who have taken RASCAL training for the many helpful comments that they provided to us.

1 Nuclear Power Plant Source Term Calculations

RASCAL 3.0.5 has a module called "Source Term to Dose" that first calculates a time-dependent "source term," for the atmospheric release of radioactive materials. The source term is the rate at which radioactive material is released from the facility into the atmosphere. The code next calculates the atmospheric transport and dispersion and the deposition of the radioactive materials that were released. Finally, the code calculates doses from the cloud shine, ground shine, and inhalation pathways as a function of location.

This chapter describes how the "Source Term to Dose" module in RASCAL 3.0.5 calculates the time-dependent source term for nuclear power plant accidents. The methods used in the RASCAL 3.0.5 source term calculations for nuclear power plant accidents are based largely on the methods described in NUREG-1228 (McKenna and Giitter 1988).

Before we describe the detailed methods that RASCAL 3.0.5 uses to calculate specific a source term types, we first describe the nuclear power plant parameters that are used in the calculations.

1.1 Nuclear Power Plant Parameters

1.1.1 Core Inventories

For nuclear power plant source terms based on core damage, the radionuclide inventories assumed to be in the reactor core are shown in Table 1.1.

The values in Table 1.1 are for low-enriched uranium fuel. The values were derived from Table 2.2 of NUREG-1228 (McKenna and Giitter 1988). The derivation was done by dividing the fission product inventories in NUREG-1228 by 3 to convert from Ci/MWe to Ci/MWt, and then rounding to 2 significant figures. The inventories of nuclides with a half-life of more than one year were scaled up from a burnup of 18,000 to 30,000 MWD/MTU.

The inventories in NUREG-1228 were derived from the Reactor Safety Study, WASH-1400. Table VI 13-1 of WASH-1400 ranked the nuclides in the core by importance to early health effects. The nuclides with greater importance for early health effects were included in the NUREG-1228 core inventory list. In additional, NUREG-1228 added noble gases that had lesser importance for early health effects because noble gases are the most likely group of fission products to be released to the environment by a nuclear power plant accident. According to Figure 2-4 of NUREG-1228, the iodine and tellurium nuclides contribute almost two-thirds of the bone marrow dose for a major nuclear power plant release. Krypton, cesium, strontium and barium are the other major contributors.

Table 1.1 Nuclear Power Plant Core Inventory During Operation for Low Enriched Uranium Fuel (30,000 MWD/MTU burnup)

Nuclide	Core Inventory Ci/MWt	Nuclide	Core Inventory Ci/MWt
Ba-140	5.30e+04	Ru-103	3.70e+04
Ce-144	2.80e+04	Ru-106	1.33e+04
Cs-134	4.17e+03	Sb-127	2.00e+03
Cs-136	1.00e+03	Sb-129	1.10e+04
Cs-137	2.67e+03	Sr-89	3.10e+04
I-131	2.80e+04	Sr-90	2.00e+03
I-132	4.00e+04	Sr-91	3.70e+04
I-133	5.70e+04	Te-129m	1.80e+03
I-134	6.30e+04	Te-131m	4.00e+03
I-135	5.00e+04	Te-132	4.00e+04
Kr-85	3.17e+02	Xe-131m	3.30e+02
Kr-85m	8.00e+03	Xe-133	5.70e+04
Kr-87	1.60e+04	Xe-133m	2.00e+03
Kr-88	2.30e+04	Xe-135	1.10e+04
La-140	5.30e+04	Xe-138	5.70e+04
Mo-99	5.30e+04	Y-91	4.00e+04
Np-239	5.50e+05		

Reference: Derived from NUREG 1228, Table 2.2 (McKenna and Giitter 1988) which in turn derived its table from WASH 1400

The inventories in Table 1.1 are based on a burnup of 30,000 MWD/MTU. RASCAL 3.0.5 adjusts the inventory of radionuclides that have a half-life exceeding one year to account for burnup. The inventory for the specified actual burnup, I_{ACTUAL}, is calculated only for nuclides with a half-life of more than one year using the Equation 1.2 below. There is no burnup adjustment for nuclides with a half-life less than one year.

$$I_{ACTUAL} = I_{30,000} \times \frac{BURNUP_{ACTUAL}}{30,000\,MWD\,/\,MTU} \tag{1.1}$$

If the reactor is shut down prior to the start of the release, the radionuclide inventories are adjusted to account for radiological decay and ingrowth. In addition, at the end of each time step, the activities of the nuclides present are adjusted to account for radiological decay and ingrowth. The minimum activity of a nuclide allowed in a source term time step is 10^{-15} Ci.

1.1.2 Coolant Inventories

RASCAL 3.0.5 uses coolant inventories for some accident types. The concentrations that RASCAL 3.0.5 uses for normal coolant are given in Table 1.2. Those normal coolant concentrations are taken from ANSI/ANS 18.1-1999. During steady-state conditions, iodine and other fission products may escape from fuel rods having clad defects and enter the reactor coolant system. Since the internal pressure in the fuel rod is balanced with the coolant pressure outside the fuel rod during steady-state conditions, the rate of escape is low. The fission products that do escape into the reactor coolant system are continually removed by the reactor coolant system purification cleanup resulting in a low equilibrium concentration.

However, if a reactor transient causes the pressure of the reactor coolant system to decrease rapidly, the escape rate from fuel rods can increase and cause a temporary increase, or "spike," in the coolant concentrations. There is also a belief that coolant water can enter fuel rods through cladding defects. If the reactor coolant system pressure suddenly decreases, this water could leach off iodine and cesium salts deposited on the inner clad surfaces, increasing the iodine and cesium available for escape during the transient.

RASCAL 3.0.5 can also calculate an inventory for "spiked" coolant. RASCAL 3.0.5 assumes that the concentrations of halogens (iodine) and alkali metals (cesium) in the coolant increase by the spiking factor. RASCAL 3.0.5 uses a default spiking factor of 100, but the user can enter a different value.

Table 1.2 Radionuclide Concentrations in Reactor Coolant

Nuclide	PWR coolant concentration	BWR coolant concentration	Nuclide	PWR coolant concentration	BWR coolant concentration
	Ci/g	Ci/g		Ci/g	Ci/g
Ba-140	1.30e-08	4.00e-10	Mo-99	6.40e-09	2.00e-09
Ce-144	4.00e-09	3.00e-12	Np-239	2.20e-09	8.00e-09
Co-58	4.60e-09	1.00e-10	Ru-103	7.50e-09	2.00e-11
Co-60	5.30e-10	2.00e-10	Ru-106	9.00e-08	3.00e-12
Cs-134	3.70e-11	3.00e-11	Sr-89	1.40e-10	1.00e-10
Cs-136	8.70e-10	2.00e-11	Sr-90	1.20e-11	7.00e-12
Cs-137	5.30e-11	8.00e-11	Sr-91	9.60e-10	4.00e-09
H-3	1.00e-06	1.00e-08	Tc-99m	4.70e-09	2.00e-09
I-131	2.00e-09	2.20e-09	Te-129m	1.90e-10	4.00e-11
I-132	6.00e-08	2.20e-08	Te-131m	1.50e-09	1.00e-10
I-133	2.60e-08	1.50e-08	Te-132	1.70e-09	1.00e-11
I-134	1.00e-07	4.30e-08	Xe-131m	7.30e-07	0
I-135	5.50e-08	2.20e-08	Xe-133	2.90e-08	0
Kr-85	4.30e-07	0	Xe-133m	7.00e-08	0
Kr-85m	1.60e-07	0	Xe-135	6.70e-08	0
Kr-87	1.70e-08	0	Xe-138	6.10e-08	0
Kr-88	1.80e-08	0	Y-91	5.20e-12	4.00e-11
La-140	2.50e-08	4.00e-10			
Mn-54	1.60e-09	3.50e-11			

Reference: ANSI/ANS 18.1 1999.

1.1.3 Reactor Coolant System Water Mass

The mass of water in the reactor coolant system for each plant is stored in the RASCAL 3.0.5 facility database. The values are in kilograms of water. The values are converted into gallons for display to the user by the user interface.

The RCS coolant system water masses were estimated for each reactor. The document ANSI/ANS 18.1 *Radioactive Source Term for Normal Operation of Light Water Reactors* provided the following information:

Reactor	Mass of water in reactor vessel (BWR) or RCS (PWR)	Source table
BWR reference at 3,400 MWt	1.7e5 kg	1
PWR - with u-tube SG reference at 3,400 MWt	2.5e5 kg	2
PWR with once-thru SG reference at 3,400 MWt	2.5e5 kg	3

For each reactor, the actual licensed power (MWt) was divided by the reference power (3,400 MWt), then multiplied by the mass of the water (kg from above table) to estimate the coolant mass.

For example, Beaver Valley Unit 1 (a PWR) had a licensed power of 2,652 MWt. Thus:

$$(2652 / 3400) \times 2.5e5 = 1.95e5 \text{ kg of water in the RCS}$$

This method produces only the approximate mass. More accurate site-specific values could be obtained from plant technical specifications, but the improved accuracy was not thought to be worth the effort.

1.1.4 Reactor Containment Volumes

The containment volumes in the RASCAL 3.0.5 database were taken from NUREG/CR-5640 (Lobner, Donahoe, and Cavallin, 1990). For PWRs, the volumes are the total containment volumes. For BWRs, the volumes are the drywell volumes. Units in the database are in cubic feet.

1.1.5 Reactor Power Levels

Reactor power levels are listed in the RASCAL 3.0.5 database in units of MWt. These represent the maximum power at which the reactor is allowed to operate. This value is used as the default value for average reactor power but may be changed by the user from the RASCAL 3.0.5 user interface. These values were originally taken from the U.S. Nuclear Regulatory Commission Information Digest. They have been updated for RASCAL 3.0.5 to be current with NRC approved power upgrades as of December 2006.

Website: www.nrc.gov/reactors/operating/licensing/power-uprates.html

1.1.6 Fuel Burnup

Two burnup numbers are contained in the RASCAL 3.0.5 database. The first is the average fuel burnup (MWD/MTU) for each reactor. A value of 30,000 MWD/MTU is used in the database for all reactors. This represents a core that is roughly two-thirds of the way to end of core life assuming typical current fuel management practices. The value changes with time and with the mix of old and new fuel in the core. The value should represent an average over the entire core. The user can change the value if more information is available, but usually this will not significantly change the calculated projected doses. This burnup number is used to adjust the available inventory of radionuclides with a half-life greater than one year (see Section 1.1.1).

1-5

The second burnup number is the average burnup for spent fuel in storage. A value of 50,000 MWD/MTU is used in the database. Again, the user may change the value if a better number is available. The spent fuel burnup is used to generate source terms for spent fuel accidents using the method in Section 1.1.1.

1.1.7 Number of Assemblies in the Core

The RASCAL 3.0.5 database contains the number of fuel assemblies in each reactor core. The values are taken from NUREG/CR-5640 (Lobner, Donahoe, and Cavallin, 1990). The numbers are used only with spent fuel accident calculations when estimating the source term activity for a fuel assembly (see Chapter 2).

1.1.8 Design Pressure

A design pressure for each reactor containment is included in the RASCAL 3.0.5 database. The values are in pounds per square inch. The design pressures are taken from NUREG/CR-5640 (Lobner, Donahoe, and Cavallin, 1990). This value is not changeable by the user from the RASCAL 3.0.5 user interface. The design pressure is not used in RASCAL 3.0.5 calculations, but the user can compare the actual containment pressure with the design pressure to determine if the actual leak rate is likely to be near or below the design leak rate.

1.1.9 Design Leak Rate

A design leak rate for each reactor containment is included in the RASCAL 3.0.5 database. The values are in percent of containment volume per day at design pressure. The design leak rates are taken from NUREG/CR-5640 (Lobner, Donahoe, and Cavallin, 1990). This value is not changeable by the user from the RASCAL 3.0.5 user interface. The design leak rate is the default containment leak rate, but the RASCAL 3.0.5 user can select any other leak rate more appropriate for a particular accident.

1.2 Source Term Types

1.2.1 Basic Method to Calculate Source Terms

A source term is defined as the activities of each radionuclide released to the environment as a function of time. The basic method to calculate a source term is to divide the nuclear power plant into compartments and then calculate the activities entering the compartment and the activities being removed from the compartment during time steps of fairly short duration. The time steps generally have a 15-minute duration.

As an example, consider a loss-of-coolant accident after reactor shutdown in which fuel is damaged and radionuclides are released to the containment and then to the atmosphere. The first compartment is the fuel. RASCAL 3.0.5 will first calculate the release from the fuel to the containment atmosphere. Since the reactor is shutdown, no new fission products are being produced. The radionuclide inventory of the fuel is being depleted during each time step by radiological decay and by release to the containment. In addition, there will be ingrowth of some radionuclides in the fuel due to the radiological decay of their parents.

1-6

The second compartment is the containment atmosphere. The activity entering the containment atmosphere during a time step is the activity released from the fuel during that time step. Activity is removed from containment atmosphere during the time step by radiological decay, removal processes (for example, removal by containment sprays), and leakage to the environment.

Time steps may be of varying length. A source term time step starts whenever the user changes any of the time-dependent data or every 15 minutes, whichever occurs first. Time steps may be no less than 1 minute and must be an integral number of minutes. Before passing the source term to the atmospheric transport model, RASCAL 3.0.5 converts the source term time steps into 15-minute time steps that start on the hour because the atmospheric transport models require that regularity.

The remainder of this chapter describes in detail how RASCAL 3.0.5 calculates the time-dependent source term for various accident types.

1.2.2 Time Core Is Uncovered Source Term

Perhaps the most powerful and important source term type that RASCAL 3.0.5 calculates is based on the time that the core is uncovered. Almost all of the radioactivity at a nuclear power plant is contained in fuel rods. A large release is not possible unless many fuel rods are substantially damaged. The only way this can reasonably occur is by loss of water from the primary coolant system so that the reactor core is left uncovered by water. If a user estimates how long a reactor core will not be covered with water, RASCAL 3.0.5 can estimate the amount of core damage that will occur and from that estimate the activity of each fission product nuclide that will be released from the core.

When a RASCAL 3.0.5 user specifies how long the core is uncovered, RASCAL 3.0.5 will estimate how much core damage will occur based on the damage timings in Tables 1.3 for BWRs and 1.4 for PWRs. (Tables 1.3 and 1.4 are taken from Tables 3 12 and 3 13 in NUREG-1465, Soffer et al. 1995.) For example, if a BWR or PWR core is uncovered for 15 or 30 minutes, the estimated damage is 50% or 100% cladding failure, respectively. If a BWR core is uncovered for 1 hour, the estimated damage will be 100% cladding failure plus 33% core melt.

Table 1.3 BWR Event Timings and Fraction of Core Activity Inventory Released

Nuclide group	BWR core inventory release fraction		
	Cladding failure (gap release phase) (0.5 hr duration)	Core melt phase (in-vessel phase) (1.5 hr duration)	Post-vessel melt-through phase (ex-vessel phase) (3.0 hr duration)
Noble gases (Kr, Xe)	0.05	0.95	0
Halogens (I, Br)	0.05	0.25	0.30
Alkali metals (Cs, Rb)	0.05	0.20	0.35
Tellurium group (Te, Sb, Se)	0	0.05	0.25
Barium, strontium (Ba, Sr)	0	0.02	0.1
Noble metals (Ru, Rh, Pd, Mo, Tc, Co)	0	0.0025	0.0025
Cerium group (Ce, Pu, Np)	0	0.0005	0.005
Lanthanides (La, Zr, Nd, Eu, Nb, Pm, Pr, Sm, Y, Cm, Am)	0	0.0002	0.005

Reference: Table 3-12 from NUREG-1465 (Soffer et al. 1995)

The fractions shown in these tables are for the particular phase. They are not cumulative. Thus, the total fraction of core inventory released in a vessel melt-through accident is the sum of the fractions for cladding failure, core melt, and vessel melt-through.

The data in Tables 1.3 and 1.4 are the result of an expert elucidation process that considered a range of severe accident sequences. These release fractions do not envelop all potential severe accident sequences, nor do they represent any particular accident sequence. However, the timings in Tables 1.3 and 1.4 for the start of each fuel damage state was based on the accident sequence that could lead to the earliest fuel failures. The timings and release fractions in Tables 1.3 and 1.4 are essentially based on a large break loss-of-coolant accident with the reactor at full power and without the operation of emergency core cooling systems. This situation leads to very rapid uncovering of the core.

However, if there were a small break in the reactor coolant system or the emergency core cooling systems initially operated successfully, the core will remain covered while the rate of decay heat production decreases. At lower decay heat production rates, the duration of each release phase is likely to increase and the release fractions during each release phase may well be overestimated by Tables 1.3 and 1.4. However, RASCAL 3.0.5 does not adjust its releases to account for that situation. Users of the time-core-is-uncovered source term option should understand that RASCAL 3.0.5 is likely to overestimate the speed and magnitude of the release and thus also overestimate the projected radiological doses. RASCAL 3.0.5 users should inform decisionmakers of that fact.

Table 1.4 PWR Event Timings and Fraction of Core Activity Inventory Released

Nuclide group	PWR core inventory release fraction		
	Cladding failure (gap release phase) (0.5 hr duration)	Core melt phase (in-vessel phase) (1.3 hr duration)	Post-vessel melt-through phase (ex-vessel phase) (2.0 hr duration)
Noble gases (Kr, Xe)	0.05	0.95	0
Halogens (I, Br)	0.05	0.35	0.25
Alkali metals (Cs, Rb)	0.05	0.25	0.35
Tellurium group (Te, Sb, Se)	0	0.05	0.25
Barium, strontium (Ba, Sr)	0	0.02	0.1
Noble metals (Ru, Rh, Pd, Mo, Tc, Co)	0	0.0025	0.0025
Cerium group (Ce, Pu, Np)	0	0.0005	0.005
Lanthanides (La, Zr, Nd, Eu, Nb, Pm, Pr, Sm, Y, Cm, Am)	0	0.0002	0.005

Reference: Table 3 13 from NUREG 1465 (Soffer et al. 1995)

For PWRs, the time the core is uncovered should be the time that the coolant drops below the top of the active fuel. At this level cladding failure will begin. The gap activity in each fuel rod is released suddenly when the cladding fails at some location due to overpressure. The rods near the center of the core will fail earliest with additional rods failing as the core continues to heat. This process takes about half an hour.

For BWRs, the cladding damage does not start until the water uncovers about 1/3 of the way down the fuel element. Prior to that time boiling water below will keep the fuel cool enough to prevent melting of the cladding.

For calculations using the time core is uncovered source term type, RASCAL 3.0.5 will first calculate the activity released from the fuel to either the containment atmosphere or to the coolant as appropriate for the release pathway that the user has selected. The equation is

$$A_i(k) = I_i \, AF_i(k) \tag{1.2}$$

where

I_i = the core inventory of radionuclide i
$AF_i(k)$ = the available fraction of the inventory of radionuclide i available for release from the fuel during time step k

To illustrate how $AF_i(k)$ is calculated, consider a PWR for the fourth 15-minute time step (45 minutes to 60 minutes), during which time we are entirely in the core melt phase. During the core melt phase, 95%

of the noble gases would be released over 1.3 hours according to Table 1.4. The available fraction for the release of noble gas activity from the fuel to the containment, $AF_{ng}(4)$, during that fourth 0.25 hour duration time step would be:

$$AF_{ng}(4) = 0.25 hr \frac{0.95}{1.3 hr} \quad , \tag{1.3}$$

If the user enters a time at which the core is recovered with water, core damage is assumed to stop and the release of material from the core stops at that time.

1.2.3 Ultimate Core Damage State Source Term

The user can specify the maximum damage that is expected to occur by selecting a core-damage state directly. The state that is selected will establish the source term. The user can select: normal coolant activity, spiked coolant activity, or 1 to 100% cladding failure.

The user also selects the time at which the maximum damage is expected to occur. For example, if the user believes that a maximum damage of 10% cladding failure may occur, the user enters the time at which he believes 10% of the cladding will have failed.

Normal Coolant

For normal coolant releases, RASCAL 3.0.5 uses the coolant concentrations from Table 1.2 decayed from the time of shutdown to the time entered as the point of maximum damage. The reactor coolant system inventory I_i is the concentration of radionuclide i times the total coolant mass. The available fraction for release is the mass of coolant escaping during the time step divided by the total coolant mass.

Spiked Coolant

Spiked coolant may be seen following reactor shutdown, startup, rapid power change, and reactor coolant system depressurization. Rapid increases in the iodine and other fission-product concentrations in the coolant as high as 3 orders of magnitude may occur. The default spiking factor is 100, but the user can select a spiking factor from 1 to 1000.

For spiked coolant releases, RASCAL 3.0.5 uses the coolant concentrations (Table 1.2). The concentration of all halogens (iodine) and alkali metals (cesium) in the coolant are multiplied by the spiking factor.

For both normal and spiked coolant releases, only the steam generator and containment bypass release pathways is available. The user must specify the mass leak rate at which coolant escapes the reactor coolant system. Generally, the leak rate can be assumed to be the same as the makeup flow needed to maintain the water level.

Cladding Failure

For cladding failure, RASCAL 3.0.5 uses available fractions (*AF*s) determined from Tables 1.3 and 1.4. Thus, for example, if the user entered 4% cladding failure for a BWR, the iodine release from the fuel would be: core inventory of iodine x 0.04 (fraction of cladding that failed) x 0.05 (the halogen available fraction for 100% cladding failure in Table 1.3).

In RASCAL 3.0.5, the user can no longer select core melt or vessel melt-through as he could in previous versions of RASCAL. For accidents proceeding to core melt, the user interface screen tells the user to use the "time core is uncovered" source term type. For accidents with core melt, the timing of the release would not be at all realistic using the ultimate core damage state source term option. In addition, while the user may have a relatively good idea of when core damage may begin, he may have less knowledge of when the maximum damage will occur. For these reasons, it is required that for accidents that are expected to proceed into core melt, the use of the "time core is uncovered" source term type should give more realistic results.

1.2.4 Containment Radiation Monitor Source Term

RASCAL 3.0.5 can use containment radiation monitor readings to estimate source terms that occur through the containment leakage release pathway. The user enters containment radiation monitor readings and the times of the readings. The entry of multiple readings allows the modeling of core damage that is progressing with time.

Figures 1.1 through 1.5 show the containment radiation monitor readings that can be expected due to radionuclides in the containment atmosphere from coolant or core damage. The figures show the calculated monitor readings at 1 hour and 24 hours after shutdown. These figures are taken from Figures A.5 through A.12 in RTM-96 (McKenna, et al. 1996). The bars in these figures represent the calculated containment radiation monitor readings for 1 to 100% of the labeled core-damage state.

The data in these figures were calculated for a reactor power of 3000 MWt. RASCAL 3.0.5 scales the monitor reading entered by the user to account for the difference in the 3000 MWt reactor power used to produce the figure and the actual reactor power. This scaled monitor reading, R, is calculated by

$$R = \frac{3000 \times MR}{Power} \quad , \tag{1.4}$$

where

MR = the containment monitor reading entered,
$Power$ = the reactor power, MWt.

To estimate a source term from the scaled monitor reading, RASCAL 3.0.5 first determines which figure should be used based on the containment type and, for BWRs, the monitor location (dry well or wet well). Next, RASCAL 3.0.5 determines if the data for containment "sprays on" or for "sprays off" should be used.

RASCAL 3.0.5 then adjusts the data in the figure for the time between shutdown and the monitor reading. If the hold-up time entered is less than 1 hour, the data for 1 hour is used without adjustment. If the holdup time is greater than 24 hours, the data for 24 hours is used without adjustment. If the holdup time is between 1 hour and 24 hours, RASCAL 3.0.5 does a linear interpolation to calculate a new 1 to

100% bar for actual holdup time (the time between the shutdown time and the time of the monitor reading).

Figures 1.1 to 1.5 show the containment monitor readings for: (1) normal coolant, (2) spiked coolant, (3) cladding failure, and (4) core melt. If a containment radiation monitor reading exceeds the value of 1% core melt, RASCAL 3.0.5 assumes that core melt has begun and uses a core melt source term.

If the containment radiation monitor reading is less than the value for 1% core melt, the core damage state is assumed to be cladding failure. The data in the figures for normal and spiked coolant are not used in RASCAL 3.0.5.

Users of the containment radiation monitor source term type should be aware of certain limitations on the calculated results:

- The figures represent typical reactor plants. Plant-specific conditions may make differences.

- The figures are appropriate for large-break loss of coolant accidents. If there is a small break the containment activity may rise very slowly at first causing RASCAL 3.0.5 to underestimate the amount of core damage that has occurred.

- Thermal stratification in the containment may effect the results. The containment atmosphere near the containment radiation monitors may not be representative of the containment atmosphere as a whole.

- The containment radiation monitor source term is a lagging indicator of core damage and cannot predict core damage that will occur in the future. Thus, it will be much later in its estimates of projected doses compared to the "time core is uncovered" source term.

Figure 1.1 PWR Containment Monitor Response

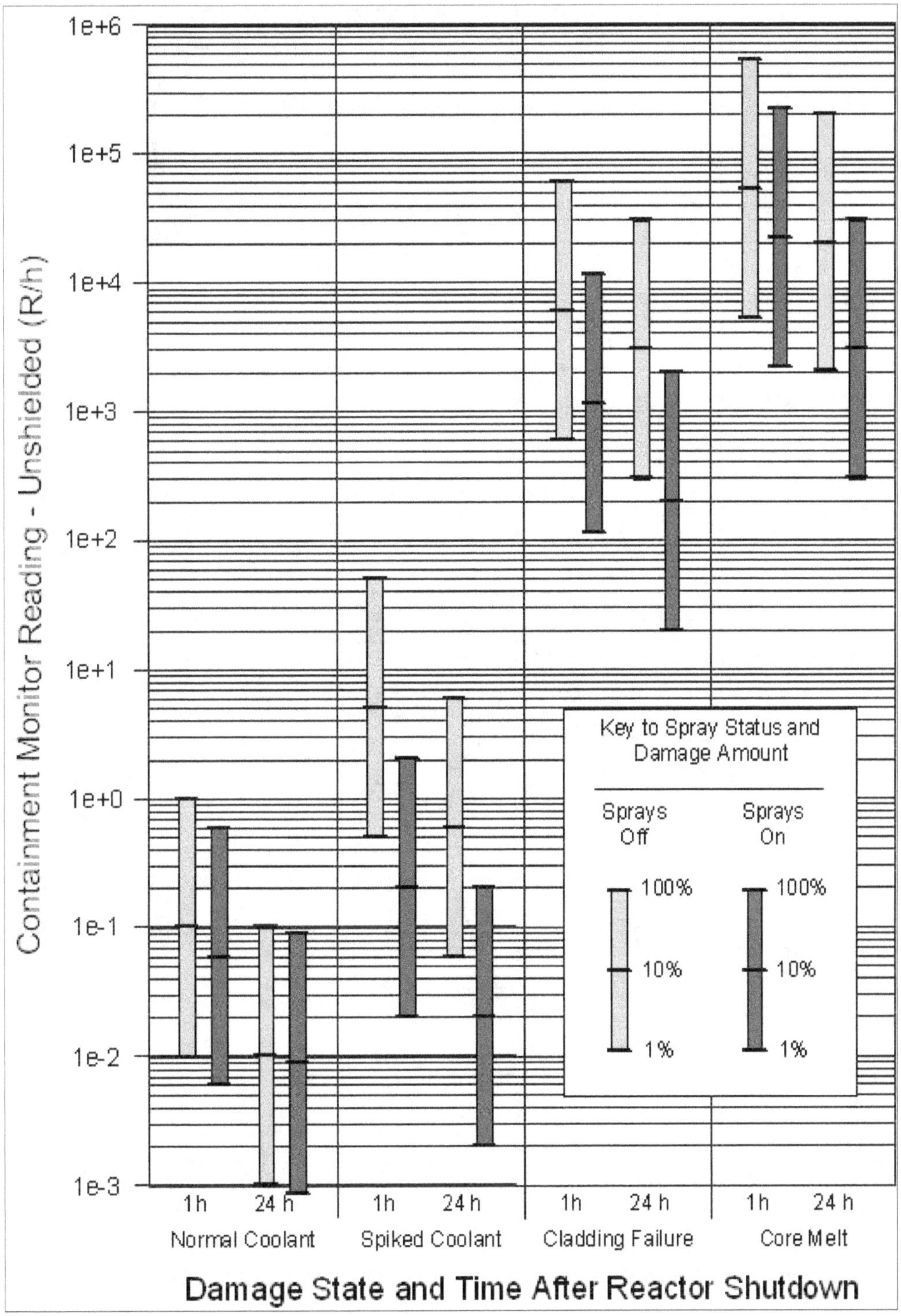

Figure 1.2 BWR Mark I and II Dry Well Containment Monitor Response

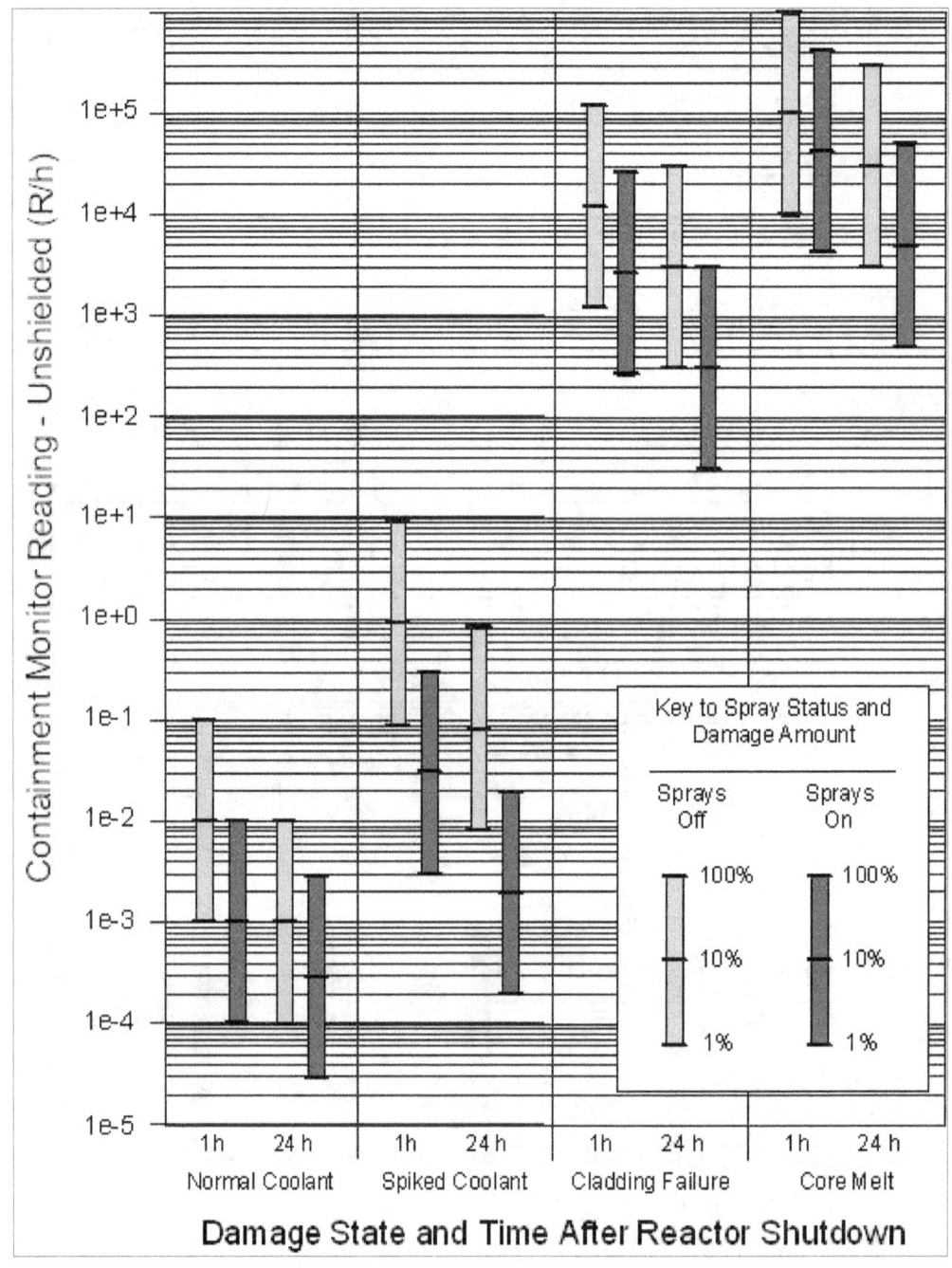

Figure 1.3 BWR Mark I and II Wet Well Containment Monitor Response

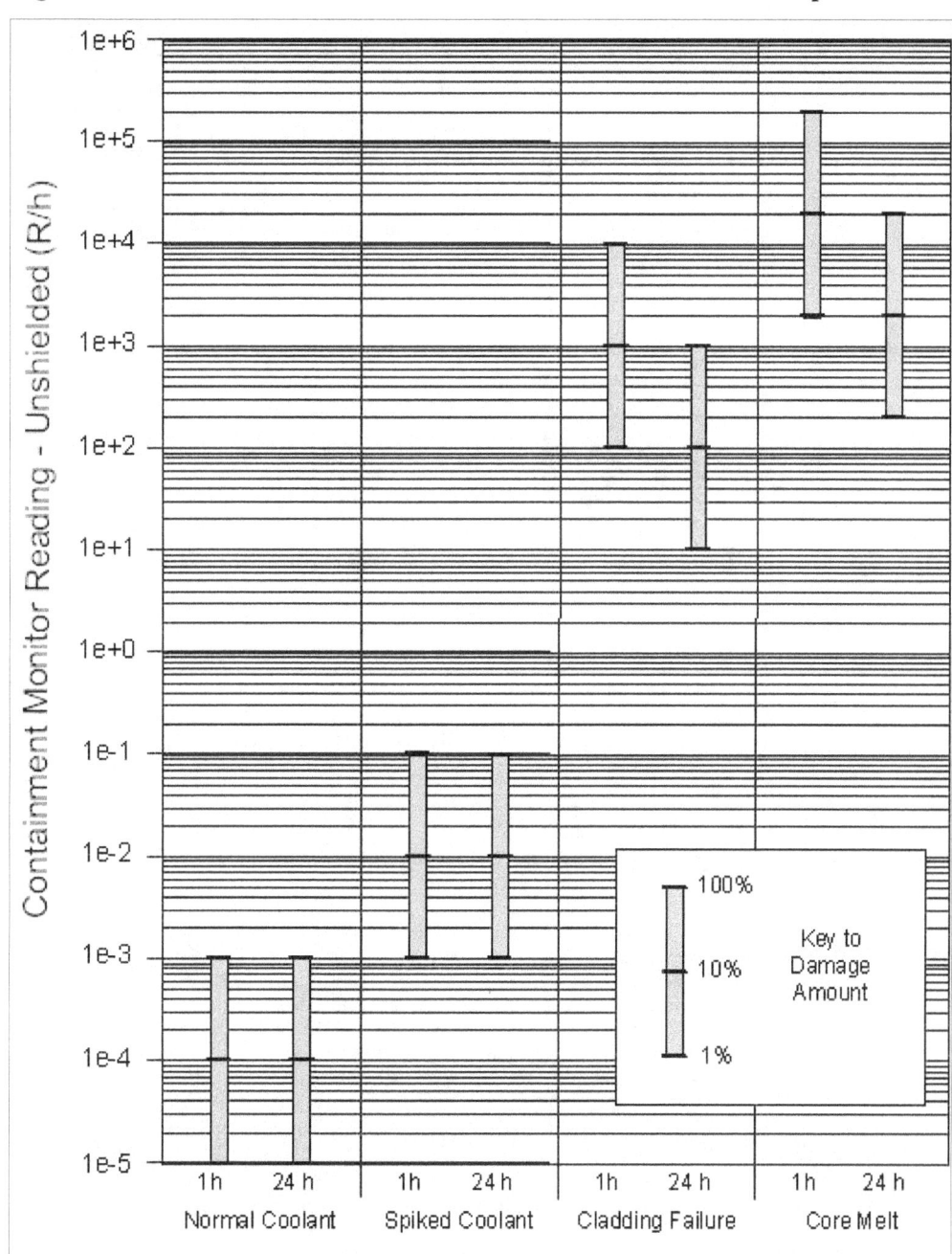

Figure 1.4 BWR Mark III Dry Well Containment Monitor Response

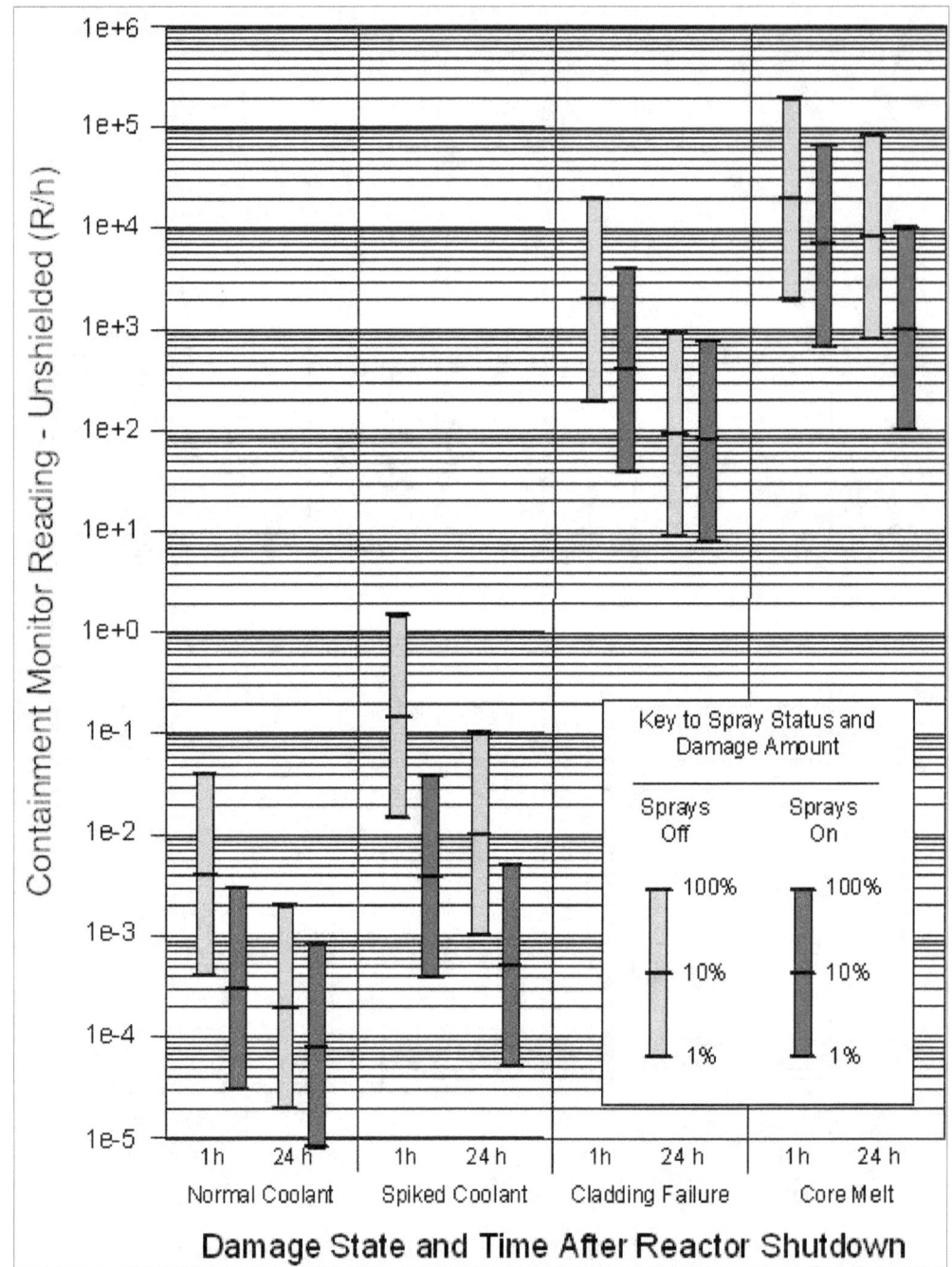

Figure 1.5 BWR Mark III Wet Well Containment Monitor Response

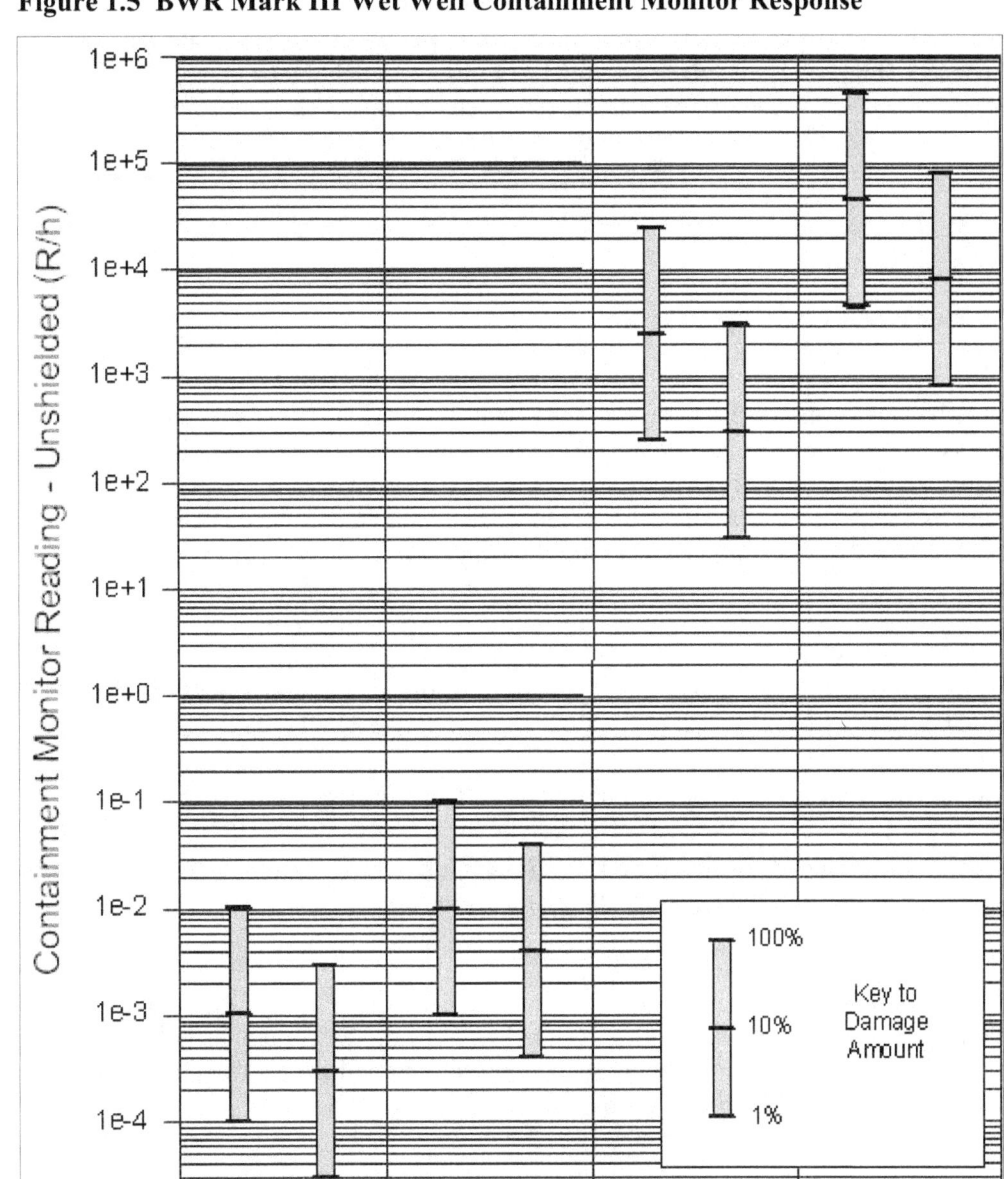

When RASCAL 3.0.5 determines that the estimated core-damage state is cladding failure or core melt, it then computes the fraction of either state that the reading represents. This percentage cannot be more than 100%. The percentage, P, of the damage state selected is calculated by

$$P = 100 \times \frac{R}{P_{1D}} \, ,$$
(1.5)

where

P_{1D} = the meter reading assumed for 1% of the core-damage state for a 3,000 MWt reactor.

The user should be very cautious in interpreting RASCAL 3.0.5 results based on containment radiation monitor readings because the calculations are subject to large uncertainties. First, the model assumes that the containment radiation monitor readings represent the full amount of damage that has occurred. However, if the fission products are delayed in entering the containment, the containment monitor readings may significantly lag behind the amount of damage that has occurred.

The model also assumes uniform mixing of fission products in the containment atmosphere. Inconsistent readings may be caused by uneven mixing in containment such as if steam rises to the top of the dome or if there is insufficient time for uniform mixing to occur. If uniform mixing has not yet occurred, the monitor readings may significantly misrepresent the amount of damage that has occurred.

The model also assumes that an unshielded monitor sees a large fraction of the containment volume. If that is not true, significant error could result. Because the mix is most likely different from that assumed in the calibration of the monitor, the actual reading at the upper end of the scale could differ significantly if a shielded detector is used for the higher radiation measurements.

1.2.5 Source Term Based on Coolant Sample

The measured concentrations of radionuclides in a nuclear power plant coolant sample can be used to define the source term when the activity being released is activity that comes from the coolant. The user must specify coolant radionuclide concentrations by nuclide.

The two release pathways that can release coolant are a steam generator tube rupture and a containment bypass pathway. The user must specify the leak rate at which coolant escapes the reactor coolant system. Generally, the leak rate can be assumed to be the same as the makeup flow needed to maintain the water level.

RASCAL 3.0.5 calculates how much activity will be released with the escaping coolant during the first 15-minute time step. For the second time step, RASCAL 3.0.5 decreases the concentration of radionuclides in the coolant to account for what has escaped. It is assumed that the makeup water being added to the primary system is clean water. Radioactive decay and ingrowth are also being accounted for to adjust the concentrations during each time step.

The reduction factors that are applied before the release to the environment are discussed later in this chapter.

1.2.6 Source Term Based on Containment Air Sample

RASCAL 3.0.5 can use the concentration of radionuclides measured in a containment air sample to define the activity released by a containment leakage pathway. (A containment air sample cannot be used to define the activity released in the steam generator tube rupture or containment bypass pathways because containment air does not exit by those pathways.)

The user enters the concentration (activity/unit volume) for each nuclide in the containment atmosphere. Multiplying the radionuclide concentrations in the containment air (activity/unit volume) times the volumetric release rate (volume/time) will equal the activity release rate (activity/time).

For PWRs, the volumes are the total containment volumes. For BWRs, the volumes are the drywell volumes. Therefore, for BWRs, RASCAL 3.0.5 analyzes only containment air samples that are taken from the drywell.

If the containment is under pressure, the density of the containment atmosphere will be greater than the density of air at normal atmospheric pressure. RASCAL 3.0.5 does not correct for this difference. The code assumes that the measurement data has been corrected for pressure. The code assumes that the data are entered in terms of activity/unit volume at the containment pressure. If the sample results are reported for the volume at atmospheric pressure, the user should increase the reported activity to account for the higher atmospheric density in the containment before the sample results are entered into the code.

1.2.7 Source Term Based on Effluent Release Rates or Concentrations

RASCAL 3.0.5 can generate a source term based on effluent measurements. The user can enter the effluent release rates (activity/unit time) by radionuclide. No radioactive decay is calculated prior to release, but radioactive decay is calculated after release. The release is assumed to be direct to the atmosphere so that reduction factors (e. g., filtering) cannot be applied.

Alternatively, the user can enter the effluent concentration (activity/unit volume) by radionuclide and the flow rate (volume/unit time). The activity of each radionuclide released to the environment is computed as the concentration times the volumetric release rate to the environment times the release duration. The effluent concentrations may be decayed over a selected time period prior to the start of release to the environment. Decay continues over the duration of the release.

Up to three sets of release rates or concentrations can be entered along with their start and end times.

1.2.8 Monitored Release - Mixtures

Nuclear power plants often report effluent mixtures of radionuclides by reporting the activities of each of three components of the mixture: noble gases, iodides, and particulates.

In RASCAL 3.0.5, a monitored mixture release may start before or after reactor shutdown. The measurement of the effluent release rate must occur during the release because if the release is not occurring, there is nothing to measure. The default for the time of measurement is at the start of the release because it is assumed to be likely that the plant operators will note the release rate as soon as it starts.

Before shutdown, the noble gas and iodine radionuclides are assumed to be in radiological equilibrium and present in the same proportion as in the core inventory shown in Table 1.1. The fraction of each noble gas nuclide in the noble gas portion of the sample is shown in Table 1.5. Radioiodines are shown in Table 1.6.

Table 1.5 Fraction of Total Noble Gas Activity for each Noble Gas Nuclide at Shutdown (burnup = 18,000 MWD/MTU)

Noble Gas Nuclides	Core activity inventory before or at shutdown, Ci/MWt (from Table 1.1)	Fraction of total noble gas activity
Kr-85	190	0.0018
Kr-85m	8,000	0.0458
Kr-87	16,000	0.0916
Kr-88	23,000	0.1317
Xe-131m	330	0.0019
Xe-133	57,000	0.3264
Xe-133m	2,000	0.0115
Xe-135	11,000	0.0630
Xe-138	57,000	0.3264

Table 1.6 Fraction of Total Radioiodine Activity for each Radioiodine Nuclide at Shutdown (burnup = 18,000 MWD/MTU)

Iodine Nuclides	Core activity inventory before or at shutdown, Ci/MWt (from Table 1.1)	Fraction of total radioiodine activity
I-131	28,000	0.1176
I-132	40,000	0.1681
I-133	57,000	0.2395
I-134	63,000	0.2647
I-135	50,000	0.2101

RASCAL 3.0.5 calculates the effluent rates for individual nuclides in a monitored mixture release using the above tables in the following manner.

If a monitored mixture release ends before shutdown, the noble gas and radioiodine are assumed to remain in the same proportions that they have in the core inventory. The activity effluent release rate for each noble gas or iodine radionuclide A_i is calculated by multiplying the total noble gas or iodine effluent rate A by the fraction for radionuclide i in Table 1.5 or Table 1.6, as appropriate.

$$A_i = A F_i \qquad (1.6)$$

where

A_i = the activity release rate of radionuclide i
A = the total measured activity release rate of noble gases or radioiodines, as appropriate
F_i = the fraction of nuclide i in the mixture from Table 1.5 or 1.6, as appropriate

Particulates are more difficult to calculate because there are so many possible radionuclides that could be released as particulates. We have taken the approach of assuming that the particulates are composed of cesium iodide (50% Cs-137 + 50% I-131). These radionuclides were selected because they are both present in the core and coolant in relatively large amounts, they are both relatively volatile and readily released from damaged fuel, and they are both biologically significant. This approach is likely to overestimate the dose from particulates, but since releases should be filtered the particulate release rate should be low and an overestimate of the dose from particulates should have little practical consequence. No radiological decay correction is done for particulates.

Case 2: Release starts before reactor shutdown and ends after reactor shutdown. Measurement is made before or at reactor shutdown.

If the sample measurement is made at or before reactor shutdown, each noble gas and iodine nuclide is assumed to be present in the monitored sample in the fractions shown in Tables 1.5 and 1.6. Thus, the release rate for each nuclide for any time step prior to shutdown is simply the effluent rate multiplied by the appropriate fraction in Table 1.5 or 1.6 as shown in the equation above.

For a time step after shutdown, the activity effluent release rate is assumed to decrease due to radiological decay. Thus, the effluent activity release rate does not remain constant, but instead decreases with time. The activity effluent release rate at or before shutdown from the equation above is corrected for radioactive decay factor as shown in the equation below. (The in-growth of daughters from the decay of the noble gases is not included in the release because they would be filtered out before release.)

$$A_i(t) = A_{0i} \exp[- \lambda_i (t - t_0)] \qquad (1.7)$$

where

$A_i(t)$ = the decay-corrected activity release rate for nuclide i at time t
A_{0i} = the activity release rate of nuclide i at time of shutdown t_0
λ_i = the radiological decay constant of nuclide i

Case 3: Release starts before or after reactor shutdown and ends after reactor shutdown. Measurement is made after reactor shutdown.

This calculation is done in three steps. First, the un-normalized fraction of each radionuclide in the mixture sample $F_i(t)$ must be determined for the sample time t_s. This is done by applying a decay correction to the fractions in Tables 1.5 and 1.6.

$$F_i(t_s) = F_{0i} \exp[-\lambda_i(t_s - t_0)] \tag{1.8}$$

where

$F_i(t_s)$ = the un-normalized decay-corrected fraction of activity for nuclide i at sample time t_s
F_{0i} = the fraction of the activity of nuclide i at time of shutdown time t_0 from Table 1.5 or 1.6
λ_i = the radiological decay constant of nuclide i

Next, the activity release rate of each radionuclide in the sample $A_i(t_s)$ is then calculated by normalizing the fractions and multiplying by the effluent release rate for that radionuclide group.

$$A_i(t_s) = A(t_s)[F_i / \sum F_i)] \tag{1.9}$$

where

$A_i(t_s)$ = the activity release rate of nuclide i at sample time t_s (after shutdown)
$A(t_s)$ = the total noble gas or radioiodine activity release rate at the sampling time t_s
ΣF_{si} = the sum of the fractions of the activities for the radionuclide group at the sampling time (to normalize the fractions for each nuclide)

The last step is to apply a correction for radiological decay. This is shown in the equation below

$$A_i(t) = A_i(t_s) \exp[-\lambda_i(t - t_s)] \tag{1.10}$$

where

$A_i(t)$ = the decay-corrected activity effluent release rate for nuclide i at time t
$A_i(t_s)$ = the activity effluent release rate of nuclide i at time of sample t_s
λ_i = the radiological decay constant of nuclide i

Note that if the time t is earlier than the sample time t_s, the exponent in the equation will be positive and the radiological decay correction will increase the activity effluent release rate to be greater than it was at the sample time..

1.3 Release Pathways

After the RASCAL 3.0.5 user has selected a source term type and entered the needed data for that source term type, he must select a release pathway to the environment. The release pathways that are available for selection will depend on the reactor type (PWR or BWR) and the source term type that the user selected.

For PWRs, there are four potential release pathways: containment leakage, containment bypass, steam generator tube ruptures, and direct to atmosphere. The release pathways available for those source term types are shown in Table 1.7.

For BWRs, there are also four potential release pathways: leakage from the drywell via the wet well, leakage through the dry well wall, containment bypass, and direct to atmosphere. The release pathways available for each source term type are shown in Table 1.8.

Table 1.7 PWR Release Pathways Available for Each Source Term Type

Source term type	Release pathway			
	Containment leakage	Containment bypass	Steam generator tube rupture	Direct to atmosphere
Time core is uncovered	X	X	X	
Ultimate core damage state - with spiked coolant release		X	X	
Ultimate core damage state - with cladding damage	X	X	X	
Containment monitor readings and containment air sample	X			
Coolant sample		X	X	
Effluent releases (rates, concentrations, and mixtures)				X

Table 1.8 BWR Release Pathways Available for Each Source Term Type

Source term type	Release pathway			
	Leakage from dry well through the wet well	Leakage from the dry well through the dry well wall	Bypass containment	Direct to atmosphere
Time core is uncovered	X	X	X	
Ultimate core damage state	X	X	X	
Containment monitor readings and containment air sample	X	X		
Coolant sample			X	
Effluent releases (rates, concentrations, and mixtures)				X

1.4 Release Pathway Models and Reduction Mechanisms

Each of the pathways listed in the previous section, except direct release to the atmosphere, will have its own characteristic potential reduction mechanisms. The reduction factors that RASCAL 3.0.5 uses are described below.

All the reduction factors are assumed to operate on all radionuclides except noble gases. None of the reduction factors reduce the activity of the noble gas release to the environment. All nuclides subject to a given reduction mechanism are assumed to have the same reduction factor. Radioiodines are treated the same as all other non-noble gas nuclides. The reduction factor multipliers are listed in Table 1.9 and described in detail in the sections below.

Table 1.9 Summary of Nuclear Power Plant Reduction Factor Multipliers

Reduction mechanism or cause	Reduction Factor Multiplier
Containment sprays (reference: NUREG/CR-4722, Figure 5)	First 0.25 h: $\exp(-12t)$ After 0.25 h: $\exp(-0.2t)$
Containment natural processes during hold-up (reference: NUREG-1150, Appendix B)	First 1.75 h: $\exp(-1.2t)$ 1.75 to 2.25 h: $\exp(-0.64t)$ After 2.25 h: $\exp(-0.15t)$
PWR Ice condenser - no fans or recirculation	0.5
PWR Ice condenser - 1 h or more recirculation	0.25
BWR release pathway from drywell via wet well with sub-cooled pool water	0.01
BWR release pathway from drywell via wet well with saturated pool water	0.05
Plate out for containment bypass pathway	0.4
Steam generator tube rupture - partitioned (break underwater)	Partitioning factor (steam concentration as fraction of SG water concentration) 0.02
Steam generator tube rupture - not partitioned (break above water level)	Partitioning factor (steam concentration as fraction of SG water concentration) 0.5
Steam generator tube rupture - condenser off gas release	0.05
Steam generator tube rupture - safety relief valve release	1
Filters	0.01
Lower limit on reduction multiplier (except for filters)	0.001
Lower limit on reduction multiplier for containment sprays (reference: NUREG/CR-4722, Figure 5)	0.03

Reference: NUREG 1228 except as noted for some specific table lines.

1.4.1 Containment Leakage in PWRs

While radionuclides are held up in the containment atmosphere, they are subject to removal from the atmosphere by water sprays and by natural processes that cause deposition on containment surfaces. If containment sprays are operating, they rapidly reduce the concentrations of all radionuclides except for noble gases. If the sprays are not operating, the natural processes such as gravitational settling and plate-out on containment surfaces by turbulent impaction gradually reduce airborne concentrations of particulates and reactive gases.

The reduction factors RF for sprays and for natural processes during holdup without sprays are both modeled as exponential functions of time t

$$RF = e^{-\lambda t} \;,$$

(1.11)

where

λ = a reduction constant for sprays or natural processes.

Both sprays and natural processes have multiple values for λ. The removal rate is larger at early times and slower at later times. Sprays and natural processes can remove particulates more readily initially and then more slowly as the readily removable particles have already been removed.

Since the user can enter release and reduction data that changes with time, it is possible to turn the sprays on and off several times. The initial spray, λ_I, applies to (1) all the activity in containment the first time the sprays are turned on, and (2) all the activity that enters the containment the first time that sprays are active. If the sprays are turned off and then turned back on, only the continuing λ_c is used. The initial λ_I for holdup applies only if the sprays were never turned on. Otherwise the continuing λ_c is used.

RASCAL 3.0.5 nuclear power plant source term calculations include a maximum effectiveness for sprays and a maximum effectiveness for all reduction, excluding filters. See Table 1.9. For each, the appropriate reduction factor or product of reduction factors computed at each time step is compared to the maximum and is not allowed to surpass it.

For PWRs with ice condenser containments, additional reductions can be taken due to interaction of the containment air with the ice. If the fans are recirculating the containment air through the ice beds for at least an hour, the activity entering the containment is reduced by using a reduction factor RF_i multiplier of 0.25. If the fans are not operating the reduction factor RF_i multiplier is 0.5. After the ice beds are exhausted, the reduction factor RF_i multiplier is 1.

1.4.2 Containment Leakage in BWRs

For BWRs, the model for reduction of radionuclides in the drywell air by sprays or natural processes is the same as for PWRs as described above. However, an additional reduction mechanism can be applied if the release from the dry well is through the wet well water.

If the release it through the wet well water and the water is sub-cooled (below the boiling point), an additional reduction factor RF_i multiplier of 0.01 is applied to all nuclides except noble gases. If the wet well water is saturated (boiling), the reduction factor RF_i multiplier is 0.05.

1.4.3 Containment Bypass

Containment bypass is a coolant release from the reactor coolant system to an auxiliary building or directly to the environment without passing through the containment atmosphere. The containment bypass release model and the reduction mechanisms are the same for PWRs and BWRs. Therefore, this section applies equally to both.

The bypass model was substantially revised in RASCAL 3.0.5 compared to previous versions. In previous models, activity escaping the fuel was transported immediately to the environment. Essentially, the previous model did not model holdup. Thus, the release rate was unrealistically fast and much too large for small leaks. The new RASCAL 3.0.5 model first distributes activity released from the fuel into the coolant with uniform concentrations. The flow rate of coolant to the environment then determines how quickly radionuclides escape to the environment.

For the bypass model, RASCAL 3.0.5 first calculates the initial concentration of each radionuclide in the coolant. If the user selects the coolant source term type, the initial coolant concentration of each radionuclide is entered directly. If the user selects the ultimate core damage state source term type, the radionuclide activity by nuclide (calculated as described in Section 1.2.3) is assumed to enter the primary coolant system. The initial concentration of each radionuclide is the activity entering the primary coolant system divided by the total coolant volume. If the user selects the time core is uncovered source term type, the initial coolant concentration is the activity released from the core during the first 15-minute time step divided by the total coolant volume.

The coolant concentrations are then multiplied by a reduction factor for plateout. The plateout multiplier for containment bypass is 0.4, which is taken from NUREG-1228. The plateout mechanism is plateout within the reactor coolant system.

The user then enters the coolant escape rate in terms of volume per unit time. Generally, the user can estimate the escape rate based on the make-up needed to maintain water levels. RASCAL 3.0.5 then calculates the activity escaping the primary coolant system during the time step by multiplying the radionuclide concentration in the coolant times the volume that escapes during the first time step.

For subsequent time steps, the concentration in the coolant is reduced to account for the activity that has escaped from the primary system. If the time core is uncovered source term type is being used, new activity enters the coolant system during each time step as described in Section 1.2.2. Thus, the coolant concentrations are being augmented during each time step by new radioactive material entering the coolant.

Radioactive decay and ingrowth are also calculated during each time step.

RASCAL 3.0.5 does not calculate any holdup or plateout in any secondary structure such as auxiliary building. However, release to the environment can be reduced by filters, if applicable.

1.4.4 Steam Generator Tube Ruptures in PWRs

The steam generator tube rupture model in RASCAL 3.0.5 is substantially changed from the model used in previous versions of RASCAL. In previous RASCAL versions, the steam generator partitioning factor was treated as a removal factor. In the new RASCAL 3.0.5, the radionuclides that enter the steam generator are truly partitioned. Therefore, there can be buildup of those radionuclides in the steam generator. The result is that the new RASCAL model will predict larger releases if the release duration is very long.

RASCAL 3.0.5 calculates the activity concentration in the primary coolant system the same way it calculated the concentration for containment bypass described in Section 1.4.3 above. The activity escaping the primary coolant system and entering the steam generator is also calculated by the same method as for bypass accidents except that no reduction factor for plateout is used.

As with the bypass release path, the RASCAL 3.0.5 user specifies the flow rate from the primary coolant system to the secondary system, which can perhaps be estimated from the makeup flow needed to maintain the water level in the primary system. The default flow rate into the steam generator is 500 gallons/minute, which is considered equivalent to the rupture of one tube in one of the steam generators.

For U-tube steam generators, the RASCAL 3.0.5 user specifies whether the tube rupture is above or below the water level in the steam generator. For once-through steam generators, the rupture is always assumed to be above the water level.

If the break is below the water level, the activity entering the steam generator is assumed to be evenly mixed in the steam generator water. The initial activity concentration (Ci/lb) in the steam generator is the activity that entered the steam generator during the first time step divided by the weight of the steam generator water. The default weight of water in a steam generator is 93,000 lbs, but the user can change the value if better information is available.

The activity concentration for non-noble gases in the steam that exits the steam generator is assumed to be the concentration in the steam generator water times a partition factor. If the rupture is below the water level, the partition factor is 50. In other words, the concentration of a non-noble gas radionuclide in the steam is assumed to be one-fiftieth (0.02) of the concentration in the steam generator water. If the break is above the water level, the partition factor is 2. In other words, the concentration in the steam is half the concentration in the water.

Note that partition factors are hold-up factors, not removal factors. The partition factors slow the release of radionuclides from the steam generator, but do not prevent it. As long as the steam generator is not isolated, the steam will continue to remove radionuclides from the steam generator water. The removal rate for the steam is the concentration in the steam times the flow rate of steam. The default for the steaming rate in a steam generator is 75,000 lbs/hr, but the user can change this value if better information is available.

Each time step, RASCAL 3.0.5 recalculates the concentration of radionuclides in the steam generator water by subtracting the activity removed in the steam during the previous time step and by adding any new activity entering from the primary coolant system through the rupture.

There are two paths by which the radionuclides in the steam can escape to the environment. The first is the safety relief valve and the other is the condenser off-gas exhaust (or steam-jet air ejector in some plants).

RASCAL 3.0.5 assumes that there is no removal of radionuclides as the steam exits through the safety release valve. If the exit is through the condenser off-gas exhaust, RASCAL 3.0.5 assumes that filters remove 95% of the non-noble gas radionuclides (multiplies activity by 0.05).

1.5 Leakage Fractions

Four methods for specifying leakage fractions for release to the environment are available in RASCAL 3.0.5: (1) specifying the percent of activity present that is released per unit time, (2) specifying a containment pressure and hole size, (3) specifying a coolant flow rate (volume or mass per unit time), and (4) specifying a "direct" release, with all activity released during the selected release duration. The methods for specifying release rates available for each release pathway are shown in Table 1.10.

Table 1.10 Methods for Specifying Release Rate for each Release Pathway

Release pathway	Method for specifying release rate
Containment leakage	% of containment volume per time or Containment pressure and hole size
Containment bypass	Coolant flow rate
Steam generator tube rupture	Coolant flow rate
Monitored effluent releases	Direct to atmosphere

1.5.1 Percent Volume per Time

This release rate method releases the activity in a fixed fraction of the containment or confinement volume per unit time.

The leakage fraction LF is used to calculate the fraction of the radionuclide inventory in the containment atmosphere that is released to the environment during each 15-minute time step. At each time step the radionuclide inventory in the containment atmosphere is adjusted to account for radiological decay and ingrowth, additions to the containment atmosphere radionuclide inventory if core damage is still occurring, removal of radionuclides from the containment atmosphere by sprays or plate-out, and removal from the containment atmosphere by release to the environment.

Consider the case in which the release rate is specified to be 100%, which corresponds to total containment failure. This rate is equal to 25% per 15-minute time step. During the first time step 25% of the activity in the containment will be released. For the second time step, the activity remaining in the containment will be reduced by subtracting the activity that escaped during the first time step. Then any applicable reduction factors such as removal by containment sprays or plateout will be calculated. Then the 25% leak will be applied to the remaining activity in the containment. Because only 25% of the material in the containment can be removed each time step, there will still be some activity remaining in the containment after an hour even at a leak rate of 100%/hr.

1.5.2 Leak Rate Based on Containment Pressure and Hole Size

RASCAL 3.0.5 can calculate the leak rate through a hole in the containment if the hole size and containment pressure are known. RASCAL 3.0.5 uses Equation 6-39 from Blevins 1984 for incompressible flow through a thin square-edged orifice. The hole is assumed to be the orifice. The mass flow rate out of containment, *MFR(k)* during time step *k*, is

$$MFR(k) = C \left(\frac{\pi D^2}{4}\right) \sqrt{2\rho \left(R(k) - P_2\right) g} \qquad (1.12)$$

where

> $C = 0.63$, an experimentally measured discharge coefficient that rarely varies outside the range of $0.59 < C < 0.65$ and is dimensionless,
> D = hole diameter in inches,
> ρ = density of containment atmosphere in pounds per cubic inch,
> $P_1(k)$= pressure in containment during time step *k* in pounds per square inch,
> P_2 = atmospheric pressure in pounds per square inch,
> g = acceleration of gravity in inches per second per second to convert between pounds and a mass unit.

The leakage fraction from containment to the atmosphere during step *k*, *LF(k)*, is

$$LF(k) = \frac{MFR(k) t}{\rho V_c} , \qquad (1.13)$$

where

> t = duration of time step *k* in seconds
> V_c = the containment volume.

If the containment pressure is less than atmospheric pressure, the leak rate is zero. The code does not compute the change in containment pressure, but the user can enter changing containment pressures as the assessment proceeds.

1.5.3 Coolant Flow Rate

Containment bypass accidents are accidents in which coolant is released without going through the containment. When the coolant escapes, it is no longer pressurized. At atmospheric pressure, the coolant will flash into steam, and the radionuclides in the coolant will become airborne. The coolant mass flow rate times the radionuclide concentration will give the radionuclide release rate. Alternatively, the coolant mass flow rate divided by the total coolant volume will give the leakage fraction for the radionuclides in the coolant.

Normally, it will not be possible to measure the coolant mass flow rate directly. However, the makeup flow needed to maintain pressure or water levels in the reactor coolant system can usually be readily determined. This makeup flow can be used as an estimate of the mass flow rate for escaping coolant.

For steam generator tube rupture accidents the makeup flow rate can be used as an estimate of the coolant mass flow rate from the reactor coolant system to the steam generator. The steaming mass flow rate in the steam generator is then a measure of the rate at which water is being removed from the steam generator. The radionuclide concentration in the steam will be the concentration in the steam generator water times the appropriate partitioning factor. The concentration in the steam times the steaming mass flow rate equals the escape rate from the steam generator.

1.5.4 Direct Release to Atmosphere

The direct release to atmosphere pathway is used with the three monitored effluent release source term types: (1) activity release rate by nuclide, (2) activity release concentration by nuclide and flow rate, and (3) monitored mixtures release rate. These releases are assumed to be measured after any removal or reduction processes have acted and represent the actual release rate to the atmosphere. Therefore, no reduction mechanisms can be applied to the releases.

The user sets a start and a stop time for the release. If the source term type is activity release rate by nuclide or activity release concentration by nuclide and flow rate, the activity release rate and the composition of the effluent are assumed to be constant over the interval. If the monitored mixture source term type is selected, the release rate and composition of the effluent changes with time to account for radiological decay. This was described in Section 1.2.8.

1.6 Decay Calculations in the Source Term

Many of the source term calculations in RASCAL 3.0.5 require the calculation of radiological decay. Because the Source Term to Dose model requires that decay calculations be performed at least every 15 min, it was determined that a significant amount of calculation time could be saved by precalculating decay over that period for all of the radionuclides in RASCAL. Since the source term calculations may also require decay over longer or shorter periods, decay was also pre-computed for 1 and 5 minutes, 1 hour, and 1, 14, and 182 days. One implication of this is that decay cannot be calculated for less than 1 minute.

The decay data used in RASCAL 3.0.5 is the decay data used in creating the dose factors in *Federal Guidance Report No. 12* (USEPA 1993). The decay data file used by source term calculations was created by running a program called Chain (Eckerman et.al 2006), that was originally written by K. F. Eckerman to read the decay data used in creating the dose factors in *Federal Guidance Report No. 12* and to solve the Bateman equations.

The source term model computes decay incrementally, starting with the longest time period for which decay data have been precalculated and stored. For each time period, the activity of the parent and its daughter products are summed. The process is repeated until the correct decay time is reached. For example, assuming that 72 minute of decay are required, the subroutines find the required nuclide decay data using a binary search and then sum it for 1 hour, then for 5 minutes twice, then for 1 minute twice.

The Source Term to Dose user interface code extracts the required 5-minute and 15-minute decay data for the radionuclides in the source term and passes them to the transport and diffusion models for the calculation of decay during atmospheric transport.

The activities of the short-lived daughters are initially set equal to their parents times the branching ratio, if applicable. These are listed in Table 1.11.

Table 1.11 Short-lived Daughters Assumed to be in Equilibrium with the Parent Activity

Parent (branching ratio)	Daughter
^{44}Ti	^{44}Sc
^{68}Ge	^{68}Ga
^{88}Kr	^{88}Rb
^{99}Mo (0.876)	^{99m}Tc
^{106}Ru	^{106}Rh
^{109}Cd	^{109m}Ag
^{113}Sn	^{113m}In
^{126}Sn	^{126m}Sb ^{126}Sb
^{129m}Te (0.65)	^{129}Te
^{135}I (0.154)	^{135m}Xe
^{137}Cs (0.947)	^{137m}Ba
^{144}Ce	^{144}Pr

1.7 Quality Assurance for the Source Term Calculations

In order to demonstrate that the source term calculations were being done correctly, RASCAL 3.0.5 time dependent source terms were compared with spreadsheet calculations. RASCAL 3.0.5 generates and displays source terms by nuclide for 15-minute time steps. A set of spreadsheets was prepared to duplicate the calculations in RASCAL 3.0.5.

A spreadsheet was developed for each source term type. The equations programmed in the spreadsheets were done to represent the methods and models described in this report. As such, the spreadsheets represented our understanding of how the calculations should be correctly done. On the spreadsheets, various combinations of release pathways, reduction mechanisms, and other operating conditions were programmed.

Several nuclides were then used for each case. A noble gas was used because no reduction mechanism should operate on noble gases. Also, in each case, a non-noble gas was used to test the reduction mechanisms.

These spreadsheet results were compared to the RASCAL 3.0.5 time dependent source term results for the selected nuclides for at least 10 time steps. If the results were in agreement, I was concluded that the calculations were being done in the same way. This was taken as demonstrating that the RASCAL 3.0.5 calculations were correct because the likelihood of both being programmed incorrectly in the same way was considered extremely unlikely.

Table 1.12 lists the spreadsheets that were prepared, the source term conditions that were used, the release pathways and reduction mechanisms that were applied.

Table 1.12 Spreadsheets Prepared for Quality Assurance Testing of the RASCAL 3.0.5 Time-dependent Source Term

Source term type	Source term conditions	Release pathway	Release pathway conditions	Nuclides tested
T me core s uncovered	Uncovered 2.5 hours	Conta nment eakage	Leak rate = 1%/h Sprays on and off	I-131
	Uncovered 2.5 hours	Steam generator tube rupture	Through steam jet a r ejector Leak rate = 1 ga /m n Part t oned	I-131
	Uncovered 2.5 hours	Conta nment bypass	Leak rate = 1 ga /m n No f ters	I-131
U t mate core damage (core damage)	50% c add ng fa ure	Conta nment eakage	Leak rate = 1%/h Sprays on and off	I-131, I-133, Kr-88
	75% c add ng fa ure	Conta nment eakage	Leak rate = 10%/h Sprays on and off	I-131, I-133, Kr-88
	50% c add ng fa ure	Steam generator tube rupture	Through steam jet a r ejector Leak rate = 100 ga /m n Part t oned	I-131, I-133, Kr-88
	75% c add ng fa ure	Steam generator tube rupture	Through safety va ve Leak rate = 250 ga /m n Not part t oned	I-131, I-133, Kr-88
	50% c add ng fa ure	Conta nment bypass	Leak rate = 100 ga /m n No f ters	I-131, I-133, Kr-88
	90% c add ng fa ure	Conta nment bypass	Leak rate = 250 ga /m n F tered	I-131, I-133, Kr-88
U t mate core damage (coo ant re ease)	Increased fue p n eakage Sp k ng factor = 100	Steam generator tube rupture	Through steam jet a r ejector Leak rate = 1000 ga /m n Part t oned	I-131, I-133, Kr-88, Sr-90

Source term type	Source term conditions	Release pathway	Release pathway conditions	Nuclides tested
	Increased fuel pin leakage Spiking factor = 50	Steam generator tube rupture	Through safety valve Leak rate = 200 gal/min Not partitioned	I-131, I-133, Kr-88, Sr-90
	Increased fuel pin leakage Spiking factor = 100	Containment bypass	Leak rate = 100 gal/min Filtered	I-131, I-133, Kr-88, Sr-90
	Increased fuel pin leakage Spiking factor = 50	Containment bypass	Leak rate = 300 gal/min No filters	I-131, I-133, Kr-88, Sr-90
Coolant sample	I-131 - 500 µCi/g I-133 - 200 µCi/g Kr-88 - 1000 µCi/g Sr-90 - 600 µCi/g	Steam generator tube rupture	Through steam jet air ejector Leak rate = 1000 gal/min Partitioned	I-131, I-133, Kr-88, Sr-90
	I-131 - 500 µCi/g I-133 - 200 µCi/g Kr-88 - 1000 µCi/g Sr-90 - 600 µCi/g	Steam generator tube rupture	Through safety valve Leak rate = 500 gal/min Not partitioned	I-131, I-133, Kr-88, Sr-90
	I-131 - 300 µCi/g I-133 - 150 µCi/g Kr-88 - 650 µCi/g Sr-90 - 500 µCi/g	Containment bypass	Leak rate = 1000 gal/min No filters	I-131, I-133, Kr-88, Sr-90
	I-131 - 150 µCi/g I-133 - 300 µCi/g Kr-88 - 450 µCi/g Sr-90 - 600 µCi/g	Containment bypass	Leak rate = 400 gal/min Filtered	I-131, I-133, Kr-88, Sr-90
Containment air sample	I-131 - 45 Ci/m^3 I-133 - 40 Ci/m^3 Kr-88 - 30 Ci/m^3 Sr-90 - 35 Ci/m^3	Containment leakage	Leak rate = pressure of 45 lb/in^2 with 2 in diameter hole	I-131, I-133, Kr-88, Sr-90

Source term type	Source term conditions	Release pathway	Release pathway conditions	Nuclides tested
	I-131 - 300 C /m³ I-133 - 400 C /m³ Kr-88 - 500 C /m³ Sr-90 - 600 C /m³	Conta nment eakage	Leak rate = Tota fa ure (100%/h)	I-131, I-133, Kr-88, Sr-90
Mon tored m xtures	NG - 2000 C /s I - 4 C /s Part - 2 C /s Not shutdown	D rect	2.5 hour re ease durat on	A nob e gases and od ne from core nventory + Cs-137
	NG - 2000 C /s I - 4 C /s Part - 2 C /s Meas 15 m n before shutdown	D rect	2.5 hour re ease durat on	A nob e gases and od ne from core nventory + Cs-137

1.8 References

American National Standard Institute (ANSI/ANS). 1999. *Radioactive Source Term for Normal Operation of Light Water Reactors.* ANSI/ANS-18.1-1999, American Nuclear Society, La Grange Park, Illinois.

Blevins, Robert D. 1984. *Applied Fluid Dynamics Handbook.* Krieger Publishing Company, Malabar, FL, 1984.

Eckerman, K.F., et al. 2006. *User's Guide to the DCAL System.* ORNL/TM-2001/190. Oak Ridge National Laboratory, Oak Ridge, Tennessee.

Lobner, P., C. Donahoe, and C. Cavallin. 1990. *Overview and Comparison of U.S. Commercial Nuclear Power Plants.* NUREG/CR-5640, SIC-89/1541, U.S. Nuclear Regulatory Commission.

McKenna, T. J., and J. Giitter. 1988. *Source Term Estimation During Incident Response to Severe Nuclear Power Plant Accidents.* NUREG-1228, U.S. Nuclear Regulatory Commission.

McKenna, T. J., et al. 1996. *Response Technical Manual: RTM-96.* Vol. 1, Rev. 4, NUREG/BR-0150, U.S. Nuclear Regulatory Commission.

Sjoreen A. L., T. J. McKenna, and J. Julius. 1987. *Source Term Estimation Using MENU-TACT.* NUREG/CR-4722, U.S. Nuclear Regulatory Commission.

Soffer, L., et al. 1995. *Accident Source Terms for Light-Water Nuclear Power Plants, Final Report.* NUREG-1465, U.S. Nuclear Regulatory Commission.

U. S. Environmental Protection Agency (EPA). 1993. "External Exposure to Radionuclide in Air Water, and Soil." *Federal Guidance Report No. 12*, EPA-402-R-93-081, U.S. Environmental Protection Agency.

U.S. Nuclear Regulatory Commission (NRC). 1975. *Reactor Safety Study: An Assessment of Accident Risks in U.S. Commercial Nuclear Power Plants*, NUREG-75/014 (WASH-1400). U.S. Nuclear Regulatory Commission.

U.S. Nuclear Regulatory Commission (NRC). 1990. *Severe Accident Risks: An Assessment for Five U.S. Nuclear Power Plants.* NUREG-1150. U.S. Nuclear Regulatory Commission.

2 Spent Fuel Storage Source Term Calculations

RASCAL 3.0.5 can calculate source terms for three types of spent fuel storage accidents: (1) releases from spent fuel stored in a pool when the water drains from the pool causing the fuel to become uncovered, overheating the fuel, and causing cladding damage, (2) releases from spent fuel stored in a pool when the fuel is damaged while it is under water, and (3) releases from spent fuel in a dry storage cask when an accident causes both damage to the cladding of the fuel and loss of the integrity of the cask.

2.1 Basic Method to Calculate Spent Fuel Source Terms

The method to calculate source terms for spent fuel accidents is similar to the method for the nuclear power plant accident source terms, but it is simpler. The method is simpler because the model assumes that the entire release from the spent fuel is released instantly. This approach does not account for the time that it takes for radionuclides to escape from the damaged spent fuel, but the amount escaping is correct even though the timing is not realistic.

To perform the calculation, RASCAL 3.0.5 first calculates the activity of each radionuclide i that is present in the spent fuel (the "inventory I_i"). Second, it calculates the fraction of the inventory of each radionuclide i that is available for release from the spent fuel for the accident being evaluated, the available fraction AF_i. Third, the product of those two terms is multiplied by a reduction factor RF_i (for example, for reduction by filters). Reduction factors can include several factors working simultaneously. Last, RASCAL 3.0.5 calculates the source term by radionuclide $S_i(k)$ released to the atmosphere during time step k by multiplying by the leakage fraction $LF(k)$ for time step k. These calculations are described by the equation:

$$S_i(k) = I_i \times AF_i \times RF_i \times LF(k) \ , \tag{2.1}$$

where

    $S_i(k)$ = activity of radionuclide i released to the environment during time step k,
    I_i = inventory of radionuclide i
    AF_i = fraction for radionuclide i available for release
    RF_i = reduction factor for radionuclide i
    $LF(k)$ = leakage fraction to the environment during time step k.

2.2 Spent Fuel Radionuclide Inventories

To calculate the inventory of each radionuclide in the spent fuel at the time of the accident I_i, RASCAL 3.0.5 starts with the inventories per MWt in Table 1.1, "Nuclear Power Plant Inventory During Operation for Low Enriched Uranium Fuel (30,000 MWD/MTU)." The core inventory is calculated by multiplying by the reactor power. The default reactor power is 100% of the rated power but can be changed by the user if appropriate.

The inventories of radionuclides with a half-life of longer than one year from Table 1.1 are then adjusted for burnup using Equation 1.2. The default burnup for spent fuel is 50,000 MWD/MTU, but the user can adjust this value if desired.

RASCAL 3.0.5 then calculates the radionuclide inventories in a single fuel assembly by dividing the burnup corrected core inventory by the number of assemblies in the core (from the reactor database). These are the inventories present in a fuel assembly at the time of reactor shutdown. If the spent fuel involved in the accident is specified in terms of "batches," the batch inventories are calculated by dividing the core inventories by 3. (A batch is assumed to be one-third of a core.)

The inventories at the time of the accident are then calculated by correcting for radiological decay and ingrowth since their last irradiation.

For spent fuel assemblies damaged underwater or stored in dry casks, the user defines how long ago the fuel was removed from the reactor. A radiological decay correction using that duration is applied to all the damaged fuel. Decay and ingrowth are calculated using the methods described in Section 1.7.

A different scheme is used with the pool drained scenario. The user defines fuel amounts as either assemblies or batches in 3 age classes: less than 1 year, 1 to 2 years, and longer than 2 years. All fuel in each age category is decayed for a set time. For fuel that has been out of the reactor for 0 to 1 year, the decay time is set to one week. This simple approximation can greatly overestimate the size of the release if the fuel has been in storage much longer than one week. For fuel that has been in storage for 1 to 2 years, the decay time is set equal to 1 year. For fuel that has been out of the reactor for more than 2 years, the decay time is set equal to 2 years.

2.3 Fractions of Inventory Available for Release in Spent Fuel Accidents

The fractions of the radionuclide inventories that are available for release during an accident AF_i are shown in Table 2.1.

Table 2.1 Fuel Release Fractions Used in Spent Fuel Accidents

Nuclide group	Release fraction by release type		
	Cold gap	Hot gap	Cladding fire
Noble gases (Xe, Kr)	0.4	0.4	1
Halogens (I, Br)	3×10^{-3}	3×10^{-2}	0.7
Alkali metals (Cs, Rb)	3×10^{-3}	3×10^{-2}	0.3
Tellurium group (Te, Sb, Se)	1×10^{-4}	1×10^{-3}	6×10^{-3}
Barium, strontium (Ba, Sr)	6×10^{-7}	6×10^{-6}	6×10^{-4}
Noble metals (Ru, Rh, Pd, Mo, Tc, Co)	6×10^{-7}	6×10^{-6}	6×10^{-6}
Cerium group (Ce, Pu, Np) and Lanthanides (La, Zr, Nd, Eu, Nb, Pm, Pr, Sm, Y, Cm, Am)	6×10^{-7}	6×10^{-6}	2×10^{-6}

Source: Table 3.2 in NUREG/CR 6451 (Travis, Davis, Grove, and Azarm 1997) rounded to 1 significant figure. The cladding fire release fractions are the geometric mean of the high and low fractions.

2.3.1 Spent Fuel Pool Water Drained

Spent fuel in a spent fuel pool must remain covered with water to remove decay heat or else the fuel will heat up and the fuel cladding may be damaged. If the cladding is damaged, radioactive materials may be released from the fuel. The model in RASCAL 3.0.5 for estimating the release of radioactive materials from damaged spent fuel is based on information in NUREG/CR-6451 (Travis, Davis, Grove and Azarm 1997).

The RASCAL 3.0.5 model assumes that cladding damage will not occur until the cladding temperature exceeds 1,200 °F. If this temperature is not reached, there will be no release. When fuel recently put into storage is uncovered but the pool is not totally drained, the model assumes that the spent fuel must be uncovered for more than 2 hours in order for the fuel to reach 1,200 °F. Therefore, no release will occur if the fuel is uncovered for less than 2 hours.

The assumption of no cladding damage in less than two hours is based on the heat-up rate of 30-day-old fuel. Because it does not give credit for heat removal by steam cooling, this assumption is usually conservative.

If the spent fuel has been out of the reactor long enough, the fuel may never reach 1,200 °F, the temperature for cladding damage. The RASCAL 3.0.5 model assumes that the temperature of the fuel will not become high enough to cause cladding damage if PWR fuel that has been out of the reactor for more than 1 year or if BWR fuel that has been out of the reactor for more than 0.5 year. This is based on the observation that one-year-old fuel PWR required 10 hours or more to reach 1,200 °F. With steam cooling this PWR fuel is not expected to reach this temperature.

In summary, there will be no release under the following conditions:

▸ The pool is not totally drained so that steam cooling can occur **and**

- the spent fuel is uncovered for less than 2 hours
  - or -
- the spent fuel has been in storage for longer than 1 year for PWR fuel or 0.5 year for BWR fuel.

▸ The pool is totally drained but for less than 2 hours **and**

- the spent fuel has been in storage for longer than 1 year for PWR fuel or 0.5 year for BWR fuel.

A release is expected if the above conditions are not met. The available fractions (AF_i) will normally be those for a hot gap release using the release fractions in Table 2.1. However, under certain circumstances, the temperature of the fuel can rise so high that a cladding fire will result. In those circumstances, the available fractions will be those in Table 2.1 for a cladding fire.

The cladding fire release fractions are used under the following conditions:

▸ The pool is totally drained for at least 2 hours during which no steam cooling occurs **and** the pool contains BWR fuel or high density racked PWR fuel less than 1 year in storage.

2-3

For PWR fuel, the user enters the number of fuel batches or assemblies that are one, two, and three years old. For BWR fuel, the user enters the number of fuel batches or assemblies that are 180 days and two years old. (No other fuel ages are allowed.) If the reactor is a PWR, the user enters whether the density of the fuel pool racking is high or low. The user enters: the length of time the fuel is uncovered, if and when the fuel was recovered, and if and when the pool was drained. When the fuel is uncovered, there is no fission product reduction due to scrubbing in the pool.

RASCAL determines which fuel release fractions to use based on conditions in the pool, the fuel racking, and the fuel age. If the pool is never totally drained, only fuel that has been in storage for one year or less is considered to be damaged and the release fractions used are those for a hot-gap release. In a PWR with high-density pool racking or in a BWR, if the pool is totally drained for at least 2 hours all fuel is damaged and the release fractions used are for a fire release. In a PWR with low-density pool racking if the pool is totally drained for at least 2 hours a fire release occurs in all fuel that has been in storage for one year or less. Fuel that has been stored for more than one year is involved in a fire release only if there is at least one batch of fuel that has been in storage for one year or less. If all fuel is older than one year, then hot-gap release fractions are used for all the fuel.

2.3.2 Fuel Damaged Under Water

Fuel that is mechanically damaged under water is assumed to remain cold but experiences cladding failure. The available fractions AF_i are the cold gap release fractions in Table 2.1. All damaged fuel is assumed to have been stored for the same length of time.

2.3.3 Release from a Dry Storage Cask

A release from a dry storage cask accident may be assumed to occur in two situations: (1) an accident causes damage to the cladding of the fuel stored in the cask and also causes the integrity of the fuel cask to be lost, or (2) cooling is lost for more than 24 hours resulting in fuel heat up and cladding failure with subsequent loss of cask integrity.

For either accident type, the user defines the number of fuel assemblies involved in the accident by either selecting the type of fuel cask or by directly entering the number of assemblies. If the user specifies the cask type, the calculations can only be done for a single cask. The major fuel damage option allows the user to specify what percentage of the stored fuel has sustained the damage.

For the first accident type, mechanical damage to cladding and the cask, the assemblies would be damaged without heating. Therefore, the available fractions AF_i will be the cold gap release fractions in Table 2.1. For an accident with loss of cooling for more than 24 hours, cladding failure is assumed to occur because the temperature rise causes melting of the cladding. For this case, the hot gap release fractions in Table 2.1 are used for the AF_i.

No release is assumed to occur if cooling is lost for less than 24 hours or the cask is engulfed in a fire because the casks are designed to maintain their integrity in those situations.

2.4 Release Pathways and Reduction Factors

For all spent fuel accidents, the RASCAL 3.0.5 user specifies the times for the start and the end of the release.

For spent fuel pool accidents, the release from the spent fuel is assumed to be into a building. If the release pathway to the environment passes through filters, a reduction factor of 0.01 is applied to all radionuclides except noble gases.

For spent fuel damaged underwater in a spent fuel pool, a reduction factor of 0.01 is applied to all radionuclides except noble gases to account for scrubbing by the water in the pool. This factor is in additional to reduction by building filters.

For dry cask storage accidents, there are no reduction mechanisms to reduce the amount of activity released.

2.5 Leakage Fractions

For all spent fuel accidents, the user specifies the leak rate to the environment in terms of %/hour with a maximum rate of 100%/hour. For fuel casks stored outdoors, the user would normally select 100%/hour to indicate a very fast transfer rate to the environment.

2.6 References

Travis, R. J., R. E. Davis, E. J. Grove, and M. A. Azarm. 1997. *A Safety and Regulatory Assessment of Generic BWR and PWR Permanently Shutdown Nuclear Power Plants*. NUREG/CR-6451, BNL-NUREG-52498, U.S. Nuclear Regulatory Commission.

3 Fuel Cycle and Materials Source Terms

The source term calculations in RASCAL estimate the amount of radioactive (or hazardous) material released based on a wide variety of potential radiological accident scenarios. The source term calculations performed that pertain to fuel-cycle facility and materials accidents can be generally categorized as (1) fuel-cycle facility/UF_6 accidents, (2) uranium fires and explosions, (3) criticality accidents, and (4) isotopic releases (e.g., transportation, materials).

3.1 Basic Method to Calculate the Source Term

RASCAL 3.0.5 calculates source terms by time steps. For each time step, RASCAL 3.0.5 first calculates the activity that is present (the "inventory I"). Second, RASCAL 3.0.5 calculates the fraction of the inventory that is released from the reactor fuel FRF. Third, the product of those two terms is multiplied by a reduction factor RF (for example, for reduction by filters or containment sprays). Reduction factors can include several factors working simultaneously. Last, RASCAL 3.0.5 calculates the source term S released to the atmosphere by multiplying by the leakage fraction LF. These calculations are described by the equation:

$$S_i(k) = I_i(k) \times AF_i(k) \times RF_i(k) \times LF(k) , \qquad (3.1)$$

where

$S_i(k)$ = activity of radionuclide i released to the environment during time step k,
$I_i(k)$ = inventory of radionuclide i at time step k,
$AF_i(k)$ = available fraction for release for radionuclide i during time step k,
$RF_i(k)$ = reduction factor for radionuclide i during time step k,
$LF(k)$ = leakage fraction to the environment during time step k.

The time steps may be of varying length. A source term time step starts whenever the user changes any of the time-dependent data or every 15 minutes, whichever is less. Time steps may be no less than 1 minute and must be an integral number of minutes. Before passing the source term to the atmospheric transport model, RASCAL 3.0.5 converts the source term time steps into 15-minute time steps that start on the hour because the atmospheric transport model requires that regularity.

3.2 UF_6 Releases from Cylinders

Starting Inventory

The inventory (I) of UF_6 available for release from cylinders can be described by the user in two ways:

1. The user specifies the number and types of cylinders releasing their contents. Table 3.1 lists the cylinder types available and the mass of UF_6 each contains. Each cylinder is assumed to be filled to its capacity. The total starting inventory is the sum of the number of each type of cylinder times the amount of UF_6 in that type of cylinder.

2. The user specifies a total mass of UF_6 in cylinders that can be released.

Table 3.1 UF$_6$ Cylinder Inventories

Cylinder type	Available inventory of UF6 (kg)
2.5 ton (30A, 30B)	2,277
10 ton (48A, 48X)	9,539
14 ton (48Y, 48G, 48F, 48H)	12,338

Release fractions and release rates

The user next selects the form of the UF$_6$ (liquid, solid, or solid in a fire) and the type of cylinder damage leading to the release (rupture or valve/pigtail failure). Specifying these serves only to set the default release fractions and release rates shown in Table 3.2.

Table 3.2 Default Release Fractions and Rates Based on UF$_6$ Form and Cylinder Damage Type

	Cylinder rupture		Valve or pigtail failure	
Form of UF6	Release fraction	Release rate (kg/s)	Release fraction	Release rate (kg/s)
Liquid	0.65	32	see table 3.3	4
Solid	1.0	0	1.0	0
Solid in fire	1.0	8	1.0	1

Source: RTM Supplements for Paducah and Portsmouth Gaseous Diffusion Plants, 1997.

There is one special case where the UF$_6$ is liquid and the release is caused by a valve or pigtail failure. In this case, the valve location is used to set the maximum release fraction. The relationship between valve position and release fraction is shown in Table 3.3. (All cylinders are assumed to have the same maximum release fraction and release rate.)

Table 3.3 UF$_6$ Release Fractions Based on Valve Location

Valve position	Maximum release fraction
360° — top	0.3870
270° — side	0.5528
180° — bottom	0.9222

Source: The release fractions were computed using data taken from Table 22 of Williams (1995, NUREG/CR 4360).

The mass of UF$_6$ available for release is the starting inventory (I) times the available fraction (AF).

Release pathways

RASCAL 3.0.5 has three possible release pathways for UF_6 cylinder releases: direct to the atmosphere, through a building, or through filters. For some pathways, it is assumed that the UF_6 will be fully converted to HF and UO_2F_2 before entering the atmosphere. Table 3.4 lists the situations where the UF_6 is converted before release.

Table 3.4 Pathways with/without UF6 Conversion Prior to Release to Atmosphere

Pathway	No conversion of UF_6	Complete conversion to HF & UO_2F_2
Direct		
- liquid	X	
- solid		X
- solid in a fire	X	
Through building		X
Through filters		X

UF_6 is converted as 1 kg of UF_6 = 0.88 kg UO_2F_2 + 0.23 kg HF. To determine the uranium activity from the uranium mass and enrichment, RASCAL converts the mass of UF_6 to activity using the enrichment level and the specific activity. This conversion described in Section 3.9.

The direct to atmosphere pathway assumes that all material released from the cylinder(s) enters the atmosphere without being acted upon by any reduction mechanisms. The leak rate to the atmosphere is the leak rate from the cylinder(s). The UF_6 available for release is divided by the release rate to determine how many time steps are required for the release to complete. The UF_6 is released at a constant rate until exhausted.

Releases through a building or filters both allow the user to specify a release fraction for the HF and the UO_2F_2, a building air exchange rate (changes per hour), and a start and end time for the release. This release fraction is different from that described earlier. These numbers represent a reduction due to the building or filters; the previous was a reduction by the cylinder.

No radiological decay is computed in UF_6 accident scenarios.

3.3 UF_6 Releases from Cascade Systems

The UF_6 cascade release accident type is available only for the Portsmouth and Paducah gaseous diffusion plants. In addition, the cascade release source term option is available only for certain buildings of those facilities. Tables 3.5 and 3.6 list the building names and information about the default inventories and release rates.

Table 3.5 Paducah GDP Buildings and Default Inventory and Release Rates

Building name	Cells per unit	Number of units	Avg cell inventory (lbs)	Release rate (lbs/sec)
C-331	10	4	4,400	130
C-333	10	6	9,500	130
C-335	10	4	4,600	130
C-337	10	6	8,400	130
C-310	10	1	150	130

Source: RTM Supplement for Paducah Gaseous Diffusion Plant, 1997.

Table 3.6 Portsmouth GDP Buildings and Default Inventory and Release Rates

Building name	Cells per unit	Number of units	Avg cell inventory (lbs)	Release rate (lbs/sec)
X-326	20	2.5	1,000	130
X-330	10	11	5,000	130
X-333	10	8	5,000	130

Source: RTM Supplement for Portsmouth Gaseous Diffusion Plant, 1997.

Starting inventory

This starting UF_6 inventory (I) may be entered: (1) directly as a total mass of UF_6 available for release or, (2) as the mass of UF_6 per cell and the number of units or cells in the cascade that are involved in the release. When the number of cells is entered, the amount of material available is the product of the number of cells times their inventory. When the number of units is entered, the amount of material is the cell inventory time the cells per unit times the number of units.

This starting inventory (I) is multiplied by the user entered fraction available for release from the cascade to the building. This value has a default of 1.0 and represents material removed by the structure due to natural processes. The user also enters the rate at which the material escapes from the cascade into the building. Each building has a default release rate that can be changed by the user.

Release pathway

The release pathway is based on two building configurations: summer and winter.

In the summer configuration, it is assumed the building is sufficiently open to the atmosphere (hot inside with all doors and windows open) that released UF_6 has essentially an unobstructed path to the outside. This UF_6 is released to the atmosphere at the defined cascade release rate. There is no application of fractions available for release or start and end of release. Also, there is no conversion before release to HF and UO_2F_2.

In the winter configuration, the UF_6 enters the building at the defined release rate and is converted as:

1 kg of UF_6 = 0.88 kg UO_2F_2 + 0.23 kg HF.

To determine the uranium activity from the uranium mass and enrichment, RASCAL converts the mass of UF_6 to activity using the enrichment level and the specific activity. This conversion described in Section 3.9.

Releases through a building or filters both allow the user to specify a release fraction for the HF and the UO_2F_2, a building air exchange rate (changes per hour), and a start and end time for the release. This release fraction is different from that described earlier. These numbers represent a reduction due to the building or filters; the previous was a reduction by the cylinder.

No radiological decay is computed in UF_6 accident scenarios.

3.4 Fires Involving Uranium Oxide

Uranium oxide fires may occur in several different types of facilities. In the milling of uranium ore, a fire can occur in a drum of milled ore or during the process of extracting solvent. After the ore is milled, the production of reactor fuel begins with creating a powder from the UO_2. Both wet and dry processes are used to produce this powder. Uranium-oxide-contaminated waste can be stored in several forms, and any of these can be involved in a fire.

Inventory and fractions available for release

The user first selects one of five locations for the fire and specifies additional conditions. This defines the default fraction available for release (AFs) and inhalation fractions (IFs).The AFs and IFs are considered to be conservative. The IF is the fraction of the material released that is expected to be inhaled. The material is defined as all vapors and any particulate material that has a diameter of < 10 μm. (Note that the IFs are not used in the source term calculation. They are used in the calculation of inhalation dose to reduce the amount of material inhaled.) The default values for the AFs and IFs are shown in Table 3.7 (DOE 1994).

Table 3.7 Fractions Available for Release and Inhalation Fractions Used in Uranium Oxide Fires

Location of fire	Condition	AF	IF
Production process	Dry process	1×10^{-3}	1
	Wet process	3×10^{-5}	1
HEPA filter	At high temperature	1×10^{-4}	1
	Failure	1	1
Incinerator exhaust		4×10^{-1}	1
Waste fire	Solid packaged in drums	5×10^{-4}	1
	Solid loosely packed	5×10^{-2}	1
	Combustible liquid	3×10^{-2}	1
	Non-combustible liquid	2×10^{-3}	1
Uranium mill	Drum in a fire	1×10^{-3}	1
	Solvent extraction	3×10^{-2}	1

Source: DOE 1994.

Next, the user specifies the mass of the UO_2 material at risk and specifies a uranium enrichment level.

RASCAL calculates the uranium mass by first multiplying the mass of UO_2 by 0.88 (ratio of atomic weights U and UO_2). Then, the uranium mass is converted to activity based on enrichment as described in Section 3.9. The source term available for release is the product of this activity times the fraction available for release.

Release pathway

Releases outside the building have no further reductions. The release rate to the atmosphere is constant at a rate set by dividing the available activity for release (Ci) by the release duration specified.

Releases inside the building are similar but reduce the available activity to release before calculating a release rate to the atmosphere. The inventory is multiplied by a reduction factor or 0.5 for unfiltered releases and multiplied by 0.01 if filtered.

3.5 Explosions Involving Uranium Oxide

Uranium oxide explosions are characterized as (1) those caused by the detonation of high explosives in contact with the material, (2) those caused by a fire (deflagration), and (3) those caused by a sudden pressure change in the material container (venting). The UO_2 in the explosion may be in liquid, solid, or powder form, or it simply may be surface contamination.

Table 3.8 Fractions Available for Release and Inhalation Fractions Used in Uranium Oxide Explosions

Explosion characteristics	Material form of the uranium oxide	Fraction available for release	Inhalation fraction
Detonation	Liquid	1	1
	Solid	1	2×10^{-1}
	Powder	1	2×10^{-1}
	Surface contamination	1×10^{-3}	1
Deflagration	Liquid	1×10^{-6}	1
	Solid	0	0
	Powder	5×10^{-3}	3×10^{-1}
	Surface contamination	1×10^{-3}	1
Venting	Liquid	2×10^{-3}	1
	Solid	0	0
	Powder	1×10^{-1}	7×10^{-1}
	Surface contamination	1×10^{-3}	1

Source: DOE 1994.

Next, the user specifies the mass of the UO_2 material at risk and specifies a uranium enrichment level.

RASCAL calculates the uranium mass by first multiplying the mass of UO_2 by 0.88 (ratio of atomic weights U and UO_2). Then the uranium mass is converted to activity based on enrichment as described in Section 3.9. The source term available for release is the product of this activity times the release fraction.

Release pathway

Releases outside the building have no further reductions. The release rate to the atmosphere is constant at a rate set by dividing the available activity for release (Ci) by the release duration specified.

Releases inside the building are similar but reduce the available activity to release before calculating a release rate to the atmosphere. The activity is multiplied by 0.5 for unfiltered releases and multiplied by 0.01 if filtered.

3.6 Criticality Accidents

A criticality accident results from the uncontrolled release of energy from an assemblage of fissile material. In RASCAL 3.0.5, a criticality accident may be modeled using the physical system scenarios in NUREG/CR-6410 (SAIC 1998) or using criticality data entered directly by the user.

The physical systems modeled are listed in Table 3.9 along with the assumed number of fissions in the first burst and the total yield. The user selects whether to model a single or multiple bursts. The bursts are assumed to come at 10 minutes intervals and continue for eight hours (total of 48 bursts).

Table 3.9 Fission Yields Used in Criticality Calculations

System modeled in the scenario	Initial burst yield (fissions)	Total yield (fissions)
Solution <100 gal	1×10^{17}	3×10^{18}
Solution >100 gal	1×10^{18}	3×10^{19}
Liquid/powder	3×10^{20}	3×10^{20}
Liquid/metal pieces	3×10^{18}	1×10^{19}
Solid uranium	3×10^{19}	3×10^{19}
Solid plutonium	1×10^{18}	1×10^{18}
Large storage arrays below prompt critical	None	1×10^{19}
Large storage arrays above prompt critical	3×10^{22}	3×10^{22}

Source: SAIC 1998.

When using the physical systems, the number of fissions in a burst (F_B) for all except the first burst is

$$F_B = \frac{F_T - F_I}{(48 - 1)} \qquad (3.2)$$

where

F_T = the total yield (fissions) of the criticality (column 3 in Table 2.1) ,
F_I = the yield (fissions) of the initial burst (column 2 in Table 2.1).

When defined by the user, the following parameters must be set:

· number of fissions (F_I) in the first burst,
· number of fissions (F_B) in the subsequent bursts,
· burst interval in minutes, and
· duration of the criticality.

This user defined method assumes that a multi-burst event will end after 48 bursts, irrespective of the burst interval.

For both methods of defining the fission yield of the criticality, the user defines fractions available for release. The available fractions are defined for noble gases, iodines, and other nuclides which have default values of 1.0, 0.25, and 0.0005 respectively.

The user also defines shielding thicknesses for steel, concrete, and water. Those thicknesses are used to calculate the reduction in the neutron and gamma prompt shine dose due to shielding.

The assumed amounts of each radionuclide released per 10^{19} fissions are listed in Table 2.1 (SAIC 1998). These values are based on ORIGEN calculations (ORNL 1989).

To calculate the source term, RASCAL first determines the initial activity of each radionuclide present as the product of the yield of the initial burst (F_I) (in 10^{19} fissions) and the activity per 10^{19} fissions listed in Table 3.10. For each following time step, RASCAL: (1) determines if the criticality is still occurring and if enough time has passed for one or more subsequent bursts to have occurred, and if so, adds the appropriate activity as the product of the yield from these burst (F_B) and the activity per 10^{19} fissions, (2) reduces the amount of activity for the amount released, and (3) applies the release fractions and radiological decay to the result. A criticality will end when either the total number of allowed bursts have been accounted for or when the "end of criticality" time entered by the user has been reached. If the user selects a release duration that is not long enough to include all 48 bursts, the total activity released will be less than the amount listed in Table 3.10.

Table 3.10 Activity (Ci) Released in Criticality of 10^{19} Fissions

Radionuclide	Activity (Ci)	Radionuclide	Activity (Ci)
Kr-83m	1.5E2	I-131	7.3E0
Kr-85m	8.9E1	I-132	1.0E3
Kr-85	1.3E-5	I-133	1.7E2
Kr-87	1.1E3	I-134	4.2E3
Kr-88	6.6E2	I-135	5.0E2
Kr-89	4.6E4	Sr-91	3.2E2
Xe-133m	1.9E-2	Sr-92	1.2E3
Xe-133	2.7E-3	Ru-106	2.0E-2
Xe-135m	3.3E2	Cs-137	1.0E-2
Xe-135	5.2E0	Ba-139	2.4E3
Xe-137	2.4E4	Ba-140	1.1E1
Xe-138	1.0E4	Ce-143	1.0E2

Source: SAIC 1998.

Release pathway

The criticality is assumed to take place inside a building. A leak rate to the atmosphere from this building is selected from the following four choices:

100% /h (represents ordinary building ventilation)
50% /h
10% /h
4% /h (equivalent to 100% per day)

This release rate method releases a fixed fraction of the material per unit time. After the criticality stops, the release rate to the environment decreases exponentially.

The user defines a start and end of the release to the atmosphere. This describes when the radionuclide material generated by the criticality enters the environment.

The user may define reduction factors for noble gases, iodines, and other radionuclides. These are multiplied times the appropriate nuclide activities to reduce the release.

3.6.1 Prompt Shine Dose Calculation

For criticality accidents, the criticality-shine dose is computed with the source term. The shielding thicknesses entered by the user are used only in this calculation. The dose in rem, D_{crit}, at 10 ft is computed as (Hopper and Broadhead 1998)

$$D_{crit} = D_{gamma} + D_{neutron} \tag{3.3}$$

$$D_{gamma} = 1 \times 10^{-15} \times F_T \times e^{-(0.386 \times S + 0.147 \times C + 0.092 \times W)} \tag{3.4}$$

$$D_{neutron} = 1 \times 10^{-14} \times F_T \times e^{-(0.256 \times S + 0.240 \times C + 0.277 \times W)} , \tag{3.5}$$

where

F_T = the total number of fissions,
S = the thickness of steel shielding in inches,
W = the thickness of water shielding in inches,
C = the thickness of concrete shielding in inches.

Doses, D, at other distances are computed using the inverse square law as:

$$D(x) = \left(\frac{10\,ft}{x}\right)^2 \times D(10\,ft) \tag{3.6}$$

where

x = distance in feet

3.7 Sources and Material in a Fire

In a fire release, the user enters the amount of each radionuclide present. No release occurs when the fire is not burning and no other types of reduction are allowed. A fire may start and stop burning only once. The default values for these fire reduction factors are from NUREG/BR-0150 (McKenna et al. 1996). The user can select fire release fractions by element, by the form of the compound, or by entering them directly. The default fire release fractions used are shown in Tables 3.11 and 3.12. Note that the total amount of activity released also depends on the release duration entered in the isotopic release pathway form. For example, if the release duration is shorter than the duration of the fire, the amount of activity released is reduced.

Table 3.11 Fire Release Fractions by Compound Form

Form of compound	Release fraction
Noble gas	1.0
Very mobile form	1.0
Volatile or combustible compound	0.5
Carbon	0.01
Semi-volatile compound	0.01
Non-volatile compound	0.001
Uranium and Plutonium metal	0.001
Non-volatile in a flammable liquid	0.005
Non-volatile in a non-flammable liquid	0.001
Non-volatile solid	0.0001

Source: Table F 2, McKenna et al. 1996.
If the compound form is not known, the user enters the fire release fractions in Table 2.1.
The fire release fraction is the fraction of the isotope released when the material is involved in a fire; it equals the total activity released (Ci) divided by the activity involved in fire (Ci).

Table 3.12 Fire Release Fractions by Element[a]

Element[b]	Release fraction[c]	Element	Release fraction	Element	Release fraction	Element	Release fraction
H (gas)	0.5	Se	0.01	I	0.5	W	0.01
C	0.01	Kr	1.0	Xe	1.0	Ir	0.001
Na	0.01	Rb	0.01	Cs	0.01	Au	0.01
P	0.5	Sr	0.01	Ba	0.01	Hg	0.01
S	0.5	Y	0.01	La	0.01	Tl	0.01
Cl	0.5	Zr	0.01	Ce	0.01	Pb	0.01
K	0.01	Nb	0.01	Pr	0.01	Bi	0.01
Ca	0.01	Mo	0.01	Pm	0.01	Po	0.01
Sc	0.01	Tc	0.01	Sm	0.01	Ra	0.001
Ti	0.01	Ru	0.1	Eu	0.01	Ac	0.001
V	0.01	Rh	0.01	Gd	0.01	Th	0.001
Cr	0.01	Ag	0.01	Tb	0.01	Pa	0.001
Mn	0.01	Cd	0.01	Ho	0.01	U	0.001
Fe	0.01	In	0.01	Tm	0.01	Np	0.001
Co	0.001	Sn	0.01	Yb	0.01	Pu	0.001
Zn	0.01	Sb	0.01	Hf	0.01	Am	0.001
Ge	0.01	Te	0.01	Ta	0.001		

[a] Table F 3 from McKenna et al. 1996. The release fraction for ruthenium was changed from the value of 0.01 in NUREG 1140 to a value of 0.1. NUREG 1140 assumed that ruthenium was nonvolatile (McGuire 1988). However, research in NUREG/CR 6218 (Power, Kmetyk, and Schmidt 1994) indicates (in Table 5) that at high temperatures ruthenium starts to become volatile. The ruthenium release fraction of 0.1 is less than the value of 0.5 used in NUREG 1140 for compounds because ruthenium is less volatile than other volatile compounds, becoming highly volatile only at temperatures not normally reached in building fires. The carbon release fraction is appropriate for carbon compounds other than CO_2. Those compounds deliver most of the dose. The dose conversion factors used for carbon are for those carbon compounds.

[b] If the specific physical form of the nuclide is known, Table 2.1 may be used.

[c] The fire release fraction is the fraction of the isotope released when the material is involved in a fire, and equals the total activity released (Ci), users should divide by the activity involved in fire (Ci).

Fire release fractions are element specific. The fire reduction factors are shown in Tables 3.11 and 3.12 and are from NUREG-1140 (McGuire 1988).

For all types of isotopic releases, if the user selects release units in mass, rather than activity, the source term is converted to curies using the specific activity of each radionuclide. The user may specify the enrichment level for enriched uranium. The enrichment level for natural uranium is assumed to be 0.7% (McKenna, et al. 1996, Table E-5). Specific activity is computed as described in Section 3.9. For natural and enriched uranium, radiological decay and dose are calculated assuming the properties of U^{238} and U^{234}, respectively. U^{234} is used rather than U^{235} because U^{234} has a specific activity about 3 orders of magnitude higher than that of U^{235}.

3.8 Isotopic Release Rates and Concentrations

These two source term types are discussed fully in Chapter 1. They are available for all the event types of RASCAL 3.0.5 except spent fuel.

3.9 Computing Uranium Specific Activity from Enrichment

The specific activity of uranium is calculated using the user entered value for the enrichment. A cubic spline is generated using the data points given in Table 3.13. This spline is then evaluated for the given enrichment to provide the specific activity.

Table 3.13 Uranium Specific Activity for Different Enrichments

Enrichment (% ^{235}U by weight)	Specific activity (µCi/g)
0.0 (depleted)	0.4
4.0	2.4
93.0	110.0

Source: Table E 5, NUREG/BR 0105, McKenna, 1996

3.10 References

Hopper, C. M., and B. L. Broadhead. 1998. *An Updated Nuclear Criticality Slide Rule.* Vol.2, NUREG/CR-6504, ORNL/TM-13322/Vol.2, U.S. Nuclear Regulatory Commission.

McGuire, S. A. 1988. *A Regulatory Analysis on Emergency Preparedness for Fuel Cycle and Other Radioactive Material Licensees (Final Report).* NUREG-1140, U.S. Nuclear Regulatory Commission.

McKenna, T. J. and J. Giitter. 1988. *Source Term Estimation During Incident Response to Severe Nuclear Power Plant Accidents,* NUREG-1228, U.S. Nuclear Regulatory Commission.

McKenna, T. et al. 1996. *Response Technical Manual: RTM-96.* Vol. 1, Rev. 4, NUREG/BR-0150, U.S. Nuclear Regulatory Commission.

Oak Ridge National Laboratory (ORNL). 1989. *ORIGEN2 Isotope Generation and Depletion Code*, CCC-371. Oak Ridge National Laboratory, Oak Ridge, Tenn.

Powers, D. A., L. N. Kmetyk, and R. C. Schmidt. 1994. *A Review of the Technical Issue of Air Ingression During Severe Reactor Accidents*. NUREG/CR-6218, SAND-94-0731, U.S. Nuclear Regulatory Commission.

Science Applications International Corp. (SAIC). 1998. *Nuclear Fuel Cycle Facility Accident Analysis Handbook*. NUREG/CR-6410, Science Applications International Corp., Reston, VA.

U. S. Department of Energy (DOE). 1994. *Airborne Release Fractions/Rates and Respirable Fractions for Non-Reactor Nuclear Facilities: DOE Handbook*. Vol. 1, DOE-HDBK-3010094, U.S. Department of Energy.

U. S. Nuclear Regulatory Commission (NRC). 1997. *RTM-96 Supplement for the Paducah Gaseous Diffusion Plant*, prepared by Pacific Northwest National Laboratory.

U. S. Nuclear Regulatory Commission (NRC). 1997. *RTM-96 Supplement for the Portsmouth Gaseous Diffusion Plant*, prepared by Pacific Northwest National Laboratory.

Williams, W. R., 1995. *Calculational Methods for Analysis of Postulated UF6 Releases*, Vol. 1, NUREG/CR-4360, U.S. Nuclear Regulatory Commission.

4 Transport, Diffusion, and Dose Calculations

RASCAL 3.0.5 uses Gaussian models to describe the atmospheric dispersion of radioactive and chemical effluents from nuclear facilities. These models have frequently been used in licensing and emergency response calculations made by the NRC staff, [e.g., PAVAN (Bander 1982), XOQDOQ (Sagendorf, Goll, and Sandusky 1982), MESORAD (Scherpelz et al.1986; Ramsdell et al. 1988], because they quickly provide reasonable estimates of atmospheric concentrations, deposition, and doses given relatively limited information on topography and meteorology.

A straight-line Gaussian plume model is used near the release point where travel times are short and plume depletion associated with dry deposition is small. A Lagrangian-trajectory Gaussian puff model is used at longer distances where temporal or spatial variations in meteorological conditions and depletion of the plume due to dry deposition may be significant.

This chapter begins with a short theoretical derivation of Gaussian plume and puff models, and then describes the implementation of those models. The chapter concludes with a description of the dose calculations in the models.

4.1 Theoretical Bases for Gaussian Models

The derivation of the Gaussian models used to describe atmospheric dispersion is discussed in many texts. Various texts including Slade (1968), Csanady (1973), Randerson (1984), and Seinfeld (1986) provide the bases for the following discussion. They may be consulted where additional detail is desired.

Atmospheric dispersion is governed, in part, by a differential equation called the diffusion equation. With a set of assumptions that can reasonably be applied to atmospheric processes, the diffusion equation has a specific, closed-form algebraic solution that is Gaussian. In one dimension, the solution is

$$\chi(x) / Q = \frac{1}{(2\pi)^{1/2}\sigma} \exp\left[-\frac{1}{2}\left(\frac{x - x_o}{\sigma}\right)^2\right] , \qquad (4.1)$$

where

$\chi(x) =$ concentration at a distance x from the center of the concentration distribution, x_o
$Q =$ amount of material released,
$\sigma =$ dispersion parameter.

Atmospheric dispersion parameters are functions of either distance from the release point or time since release. They may also be functions of atmospheric stability and surface roughness. Numerous atmospheric dispersion experiments have been conducted to evaluate dispersion parameters and to develop methods to predict dispersion-parameter values from readily available data. A number of these experiments are described by Draxler (1984).

4.1.1 Gaussian Puff Model

Using the principle of superposition, the one-dimensional solution of the diffusion equation can be expanded to three dimensions to get the basic Gaussian puff model. In a Cartesian coordinate system with x and y axes in a horizontal plane and z in the vertical, the normalized concentration in the vicinity of the puff is

$$\chi(x,y,z)/Q = \frac{1}{(2\pi)^{3/2}\sigma_x\sigma_y\sigma_z} \exp\left[-\frac{1}{2}\left(\frac{x-x_o}{\sigma_x}\right)^2\right] \exp\left[-\frac{1}{2}\left(\frac{y-y_o}{\sigma_y}\right)^2\right] \exp\left[-\frac{1}{2}\left(\frac{z-z_o}{\sigma_z}\right)^2\right]. \quad (4.2)$$

This equation, when combined with a transport mechanism to move the center of the puff (x_o, y_o, z_o), is a simplified version of the puff model in RASCAL 3.0.5. The dispersion parameters are shown as functions of direction from the puff center. However, in most implementations of the puff model, the puff is assumed to be symmetrical in the x and y directions. Hence, x and y may be replaced by the horizontal distance r from the center of the puff.

The form of Eq. (4.2) shown is appropriate if the height of the center of the puff is such that vertical dispersion proceeds unimpeded either by the ground or by an elevated layer of the atmosphere. Unimpeded vertical dispersion is generally not the case. Typically, the earth's surface and the top of the atmospheric mixing layer are assumed to be reflective surfaces. When these assumptions are made, the vertical exponential term

$$\exp\left[-\frac{1}{2}\left(\frac{z-z}{\sigma_z}\right)^2\right]$$

is replaced by a sum of exponential terms that account for reflection. This sum is

$$\sum_{n=-\infty}^{\infty}\left\{\exp\left[-\frac{1}{2}\left(\frac{2nH-h-z}{\sigma_z}\right)\right]^2 + \exp\left[-\frac{1}{2}\left(\frac{2nH+h-z}{\sigma_z}\right)\right]^2\right\},$$

where

> H = height of the top of the mixing layer, and
> h = release height.

In practice, only a small number of terms need be considered. In RASCAL 3.0.5, as in MESORAD (Scherpelz et al. 1986; Ramsdell, et al. 1988), the summation is carried out from $n = -2$ to 2. This term can be simplified if one or more of H, h, or z equals zero. For example, if H is large compared to σ_z and z is zero, the summation may be replaced by

$$2\exp\left[-\frac{1}{2}\left(\frac{h}{\sigma_z}\right)^2\right].$$

At long down-wind distances where the vertical dispersion parameter is the same magnitude as the mixing layer thickness, the puff model can be further simplified by assuming that material is uniformly distributed in the vertical. With this last assumption, the puff model becomes

$$\chi(r)/Q = \frac{1}{2\pi\sigma^2 H} \exp\left[-\frac{1}{2}\left(\frac{r}{\sigma}\right)^2\right] , \tag{4.3}$$

where

H = mixing layer thickness.

RASCAL 3.0.5 switches to the uniformly mixed model when $\sigma_z > 1.05H$.

4.1.2 Straight-Line Gaussian Plume Models

Puff models represent plumes as a series of puffs. Concentrations at a point in the plume are calculated by adding the concentrations at the point associated with all puffs in the vicinity of the point. In effect, the puff models perform a numerical time integration of concentration as puffs pass by the point. Near a release point, the meteorological conditions may be assumed to be constant as the puff moves from the source to the receptor. If the wind speed is assumed to be much greater than zero and the point for which the concentration is to be calculated is sufficiently far down wind that the change in dispersion parameters with distance as puffs pass the point can be neglected, then the puff model can be integrated analytically to give a plume model.

Assuming that the x-axis is aligned with the mean transport direction and that the mean wind speed is u, then the average concentration during plume passage is given by

$$\chi(x,y,z) = \int_{-\infty}^{\infty} \frac{Q' F_y F_z}{(2\pi)^{3/2} \sigma_x(x)\sigma_y(x)\sigma_z(x)} \exp\left[-\frac{1}{2}\left(\frac{x-ut}{\sigma_x(x)}\right)^2\right] dt , \tag{4.4}$$

where

χ = average concentration,
Q' = release rate,
F_y, F_z = lateral and vertical exponential terms, shown above,
x = downwind distance at which χ, σ_x, σ_y and σ_z are evaluated,
u = wind speed,
t = time.

On integration, the plume model becomes

$$\chi(x,y,z)/Q' = \frac{F_y F_z}{2\pi u \sigma_y \sigma_z} , \tag{4.5}$$

which is a simplified version of the straight-line Gaussian model used in RASCAL 3.0.5.

The straight-line Gaussian plume model for ground-level releases is frequently given as

$$\chi / Q = \frac{1}{\pi u \sigma_y \sigma_z} \exp\left[-\frac{1}{2}\left(\frac{y}{\sigma_y}\right)^2\right] , \tag{4.6}$$

where F_y in Eq. (4.5) is the exponential term in Eq. (4.6).

When the release and the receptor are at ground level and H is large, the sum of exponential terms that comprise F_z has a value of 2. Hence, the constant 2 in Eq. (4.5) does not appear in Eq. (4.6).

Another assumption that deserves comment is that the meteorological conditions are assumed to be horizontally homogeneous and stationary. This means that the wind direction and speed responsible for transporting the plume from the release point to the receptor and the turbulence responsible for diffusion are assumed not to change with location throughout the model domain. It also means that the meteorological conditions do not change as a function of time during the release and time required for transport. Together, these assumptions constrain the usefulness of the straight–line plume model to estimating concentrations and doses at receptors near the release point for short-duration releases; at longer distances another model is required.

4.1.3 Treatment of Calm Winds

The straight-line Gaussian plume model in Eq. (4.5) tends to overestimate concentrations and doses during low-wind speed conditions and becomes undefined for calm wind conditions because wind speed is in the denominator. This behavior results because the derivation of the straight-line Gaussian plume model assumes that the wind speed is significantly greater than zero, thus eliminating a portion of the solution of the dispersion equation that deals with low-wind speed diffusion.

To compensate for the missing part of the solution, many straight-line models assume a wind speed of 0.5 to 1 m/s when calm winds are encountered. However, this assumption does not address the other aspect of calm winds — the lack of a well-defined wind direction. No entirely satisfactory wind direction assumption exists for calm winds for a straight-line model.

The Gaussian puff model behaves well in calm winds if the dispersion parameters are a function of time instead of travel distance. In models where the dispersion parameters are calculated as the function of travel distance, dispersion ceases during calm winds, and the material distribution remains unchanged as long as the wind is calm. In either case, deposition, depletion, exposures, and doses are calculated just as they are during windy conditions.

RASCAL 3.0.5 generally uses dispersion parameters that are a function of distance, but shifts to dispersion parameters that are a function of time when the wind speed falls below 0.5 m/s (1 mph).

The puff model does not have numerical problems with a calm wind. When a puff encounters a zero wind speed, the puff remains stationary. As a result, there is no need to change to a special model when the wind speed is low or zero. However, the puff model does change the method used to calculate dispersion parameters when the wind speed is below 0.5 m/s. The calculation of dispersion parameters during low wind speeds is discussed in Section 4.3.2.

In the plume model, when the wind speed falls below 0.5 m/s (1 mph), RASCAL 3.0.5 switches from the standard Gaussian plume model previously described to a plume model derived by Frenkiel (1953). In this model, described by Csanady (1973) and Kao (1984), the dispersion parameters of σ_x, σ_y, σ_z are assumed to be functions of along wind, cross wind, and vertical turbulence levels of σ_u, σ_v, σ_w and travel time (e.g., $\sigma_y = \sigma_v t$). With this assumption,

$$\chi/Q' = \frac{\sigma_v \exp\left(-\frac{U^2}{2\sigma_v^2}\right)}{(2\pi)^{3/2}\sigma_v\sigma_w r^2}\left[1 + \left(\frac{\pi}{2}\right)^{1/2}\frac{Ux}{\sigma_v r}\exp\left(\frac{U^2 x^2}{2\sigma_v^2 r^2}\right)erfc\left(-\frac{1}{2^{1/2}}\frac{Ux}{\sigma_v r}\right)\right] , \tag{4.7}$$

where

σ_u, σ_v, and σ_w = along wind, cross wind, and vertical turbulence measures, respectively (m/s)

r = a pseudo-diagonal distance from a point directly above the release point to the intake.

The definition of r is

$$r^2 = x^2 + \left(\frac{\sigma_u}{\sigma_v}\right)^2 y^2 + \left(\frac{\sigma_u}{\sigma_w}\right)^2 z^2 , \tag{4.8}$$

where

x = the downwind distance,
y = the cross wind distance from the plume center line,
z = the vertical distance from the plume center line.

For positions under the center line of a plume, $y = 0.0$ and $z = h + \Delta h$ where h is the release height and Δh is the plume rise, if any. Thus,

$$r^2 = x^2 + \left(\frac{\sigma_u}{\sigma_w}\right)^2 (h + \Delta h)^2 . \tag{4.9}$$

Equation (4.7) is well behaved in low-wind speed conditions and gives finite χ/Q values for calm wind (mean wind velocity = 0) as long as σ_v and σ_w are nonzero. For calm winds (wind speeds < 0.5 m/s), Eq. (4.7) has a simple form that is similar to the standard Gaussian puff model. It is

$$\chi/Q' = \frac{\sigma_v}{(2\pi)^{3/2}\sigma_v\sigma_w r^2} . \tag{4.10}$$

The crosswind position y is assumed to be zero, and the resulting χ/Q' is a function only of x and z. The χ/Q' calculated at each distance is applied in all directions.

4.1.4 Model Domains and Grids

The plume model and the puff model use different model domains. The plume model domain consists of a polar grid with receptor nodes on circles at 10° intervals at eight radial distances that may be adjusted to suit the problem at hand. The puff model domain consists of a square Cartesian grid with receptor nodes uniformly spaced throughout the domain. The polar grid has a higher node density near the release point than the Cartesian grid, and conversely, the Cartesian grid has a higher node density in the far field than the polar grid.

The sizes of the model domains are linked. If the 10-mile Cartesian grid is selected, the polar grid will have a default maximum radius of 2 miles. Similarly, the 25- and 50-mile Cartesian grids have corresponding default polar grids of 5 and 10 miles, respectively. The default radial distances for the polar grids should be satisfactory for most applications. However, the RASCAL 3.0.5 user can change the radii values if other calculational distances are more appropriate.

In general, the receptor nodes for the two grids do not coincide. This fact leads to computational differences in the doses reported in the maximum value tables in the model output for the two models for wind directions other than north, east, south, or west. The doses reported for the close-in plume model are for the plume centerline at each distance. The doses reported for puff model are the highest doses calculated at nodes at about the nominal distance -- for example, 5 miles. The node with the highest dose may or may not be on plume centerline, and may be nearer to or farther from the release point than the nominal distance. When a direct comparison of doses calculated by the two models is desired, the wind direction for the period of calculations should be north, east, south, or west. For these wind directions, both the plume model and the puff model calculate plume centerline concentrations and doses.

4.2 Transport

Atmospheric transport refers to the movement of material with the wind from the source to downwind receptors. The following two sections describe the treatment of atmospheric transport in RASCAL 3.0.5.

4.2.1 Puff Model Transport

Unlike the plume model, the puff model explicitly accounts for transit time in all calculations because the model tracks the movement of individual puffs and calculates concentrations and doses based on puff positions. As a result, dose rates calculated by the puff model may be used to estimate the time of arrival of a plume and may be compared with dose rates measured in the field. Decay and ingrowth of radionuclides and depletion of the puffs as a result of wet and dry deposition are calculated at 5-minute intervals.

The puff model differs significantly from the plume model in that neither the wind data nor the wind fields are modified to force the centers of puffs to pass directly over the receptor nodes. As a result, when the wind direction is constant, the puff model may not calculate centerline concentrations and doses. However, as time goes by and atmospheric conditions (wind direction, wind speed, stability, mixing layer thickness, and precipitation) change, the puff model will give more realistic concentration and dose patterns than the plume model. In addition, the puff model will give more realistic concentration and dose patterns than the plume model when topography modifies the winds because the wind fields used by the puff model may be modified to account for topography. The wind data used by the plume model are not modified to account for topography.

The movement of puffs is controlled by the wind at the center of the puff as it moves through the model domain. The spatial variation of winds is represented in the plume model by two-dimensional fields of vectors that give the direction and speed of puff movement. These fields are prepared by the meteorological model discussed in Section 6 and are updated at 15-minute intervals based on the available wind data.

Calculation of puff movement is a six-step process. In sequence, the steps are

1. Make an initial estimate of the direction and speed of the puff movement given the current puff position and height above ground using bilinear interpolation (Press, et al. 1986) of the vectors at the nearest nodes of the field;

2. Make an initial estimate of the puff position at the end of the period using the initial estimates of direction and speed;

3. Make a second estimate of the direction and speed of puff movement using the estimated puff position at the end of the period;

4. Make a second estimate of the puff position at the end of the period using the estimate of direction and speed from step 3;

5. Average the end points calculated in steps 2 and 4;

6. Calculate the final estimate of direction and speed of puff movement using the puff's initial position and the average end point calculated in step 5.

The actual puff movement for the period may take place in one or several steps. The step size is adjusted to ensure adequate accuracy in the integration of concentrations that takes place at receptors. Errors in the integration should be less than 5% at typical wind speeds. Larger errors may occur near the release point in high wind speed conditions, because the minimum step size is 30 seconds. These larger errors should not be a problem because plume model output should be used for receptors near the release point.

The vector fields prepared by the meteorological program are for a height of 10 meters above ground. These vectors are used for puffs that represent ground-level releases. If the actual release height is greater than 12 meters, a wind-speed profile is used to adjust the transport speed from 10 meters to the puff transport height. The profile used to adjust the wind speed considers both surface friction and atmospheric stability (see Panofsky and Dutton 1984, Sections 6.4 6.6).

4.2.2 Plume Model Transport

The RASCAL 3.0.5 plume model is a straight-line Gaussian model. As this name implies, the model assumes straight-line transport based on the wind direction at the time and place of release. The plume model rounds the wind direction to the closest 10° as it calculates the transport direction to ensure that the axis of the plume passes directly over receptors.

As is common in straight-line Gaussian models, transit time is not considered in determining when material arrives at receptors; material arrives at receptors at the time of release. As a result, dose rates calculated by the RASCAL 3.0.5 plume model cannot be used to estimate the time of arrival of a plume at a receptor and are not likely to correspond with dose rates measured in the field.

Transit time, calculated using the wind speed at the release height, is used to calculate the decay of radionuclides between the source and the receptors. It is also used to calculate depletion of material in the plume due to dry and wet deposition. Decay calculations are performed at 5-minute intervals; depletion is calculated for the full transit time.

4.3 Dispersion Parameters

In RASCAL 3.0.5, the horizontal dispersion parameters (σ_y and σ_r) are calculated using empirical curves derived from the results of a large number of dispersion experiments conducted in the 1950s and 1960s. The experiments, which were conducted over relatively flat terrain, typically involved tracer releases ranging from about 10 minutes to 1 hour in duration with ground-level concentration measurements at distances ranging from 100 meters to several kilometers. Only a few direct measurements of vertical dispersion parameters (σ_z) were made. Consequently, vertical dispersion parameters were estimated with dispersion models using measured values of the horizontal dispersion parameter and measured concentrations. Dispersion parameters have been summarized in many forms. Perhaps the best known summary is the set of dispersion parameter curves called the Pasquill Gifford curves (Gifford 1976).

4.3.1 Normal Dispersion

Regulatory guidance published by the NRC includes graphic depiction of these curves, and numerical approximations to the curves are included in many computer codes used by the NRC. In RASCAL 3.0.5, dispersion parameters are estimated using the same basic algorithms that were used in earlier RASCAL versions (Athey et al. 1993) and are used in other NRC codes including PAVAN (Bander, 1982) and XOQDOQ (Sagendorf, et al., 1982). These parameterizations have generally been attributed to Eimutis and Konicek (1972). However, the σ_y parameterization is properly attributed to Tadmor and Gur (1969), and the σ_z parameterization is properly attributed to Martin and Tikvart (1968).

The basic dispersion parameter relationships used in the NRC codes are

$$\sigma_y = a_y x^{0.9031} \tag{4.11}$$

and

$$\sigma_z = a_z(x) \cdot x^{b_z(x)} + c_z(x) \tag{4.12}$$

where

x = distance from the release point (in meters),
a_y is a function of stability class, and
a_z, b_z, and c_z are empirical values that are functions of stability class and distance.

Table 4.1 gives values for a_y, a_z, b_z, and c_z. Note that 0.9031, a_y, a_z, b_z and c_z are empirical values evaluated by fitting curves. Of these constants, 0.9031 and b_z are dimensionless, c_z has dimensions of meters, and a_y and a_z have dimensions of $m^{(1.0-0.9031)}$ and ⸱⸱⸱⸱ , respectively.

Table 4.1 Constant Values for Calculation of Atmospheric Dispersion Parameters

	Distance Range (m)	Stability Class						
		A	B	C	D	E	F	G
a_y	all x	0.3658	0.2751	0.2089	0.1471	0.1046	0.0722	0.0481
a_z	x < 100 m	0.192	0.156	0.116	0.079	0.063	0.053	0.032
	100 m < x < 1000 m	0.00066	0.0382	0.113	0.222	0.211	0.086	0.052
	1000 m < x	0.00024	0.055	0.113	1.26	6.73	18.05	10.83
b_z	x < 100 m	0.936	0.922	0.905	0.881	0.871	0.814	0.814
	100 m < x < 1000 m	1.941	1.149	0.911	0.725	0.678	0.74	0.74
	1000 m < x	2.094	1.098	0.911	0.516	0.305	0.18	0.18
c_z	x < 100 m	0.0	0.0	0.0	0.0	0.0	0.0	0.0
	100 m < x <1000 m	9.27	3.3	0.0	1.7	1.3	0.35	0.21
	1000 m < x	9.6	2.0	0.0	13.0	34.0	48.6	29.2

4.3.2 Low Wind Speed (Building Wake) Correction

Atmospheric dispersion experiments did not stop after development of the Pasquill-Gifford curves. However, the emphasis of the experiments did shift. For example, there have been experiments that examined dispersion in the vicinity of buildings and other experiments that examined dispersion under low wind-speed conditions. Ramsdell (1990) describes corrections to the basic dispersion parameter relationships in Eq. (4.11) and (4.13) that improve the dispersion parameter estimates in the vicinity of building wakes.

More recent analysis of dispersion and turbulence data (Ramsdell and Fosmire 1998) suggests that the apparent enhanced dispersion noted in the vicinity of buildings at low wind speeds in wake dispersion experiments is caused by underestimation of dispersion by the basic dispersion algorithms rather than by increased turbulence in the vicinity of buildings.

This point is illustrated in Figure 4.1 which shows the ratios of normalized concentrations (χ/Q) predicted in building wakes to observed concentration normalized to actual release rate as a function of wind speed. If the errors in the predicted values were associated with the wake, they should increase with wind speed. The fact that the errors are greatest at very low speeds and decrease with increasing wind speed indicates that the problem is underestimation of dispersion in low wind speeds. The original conclusion that the enhanced dispersion was due to building wakes appears to be incorrect.

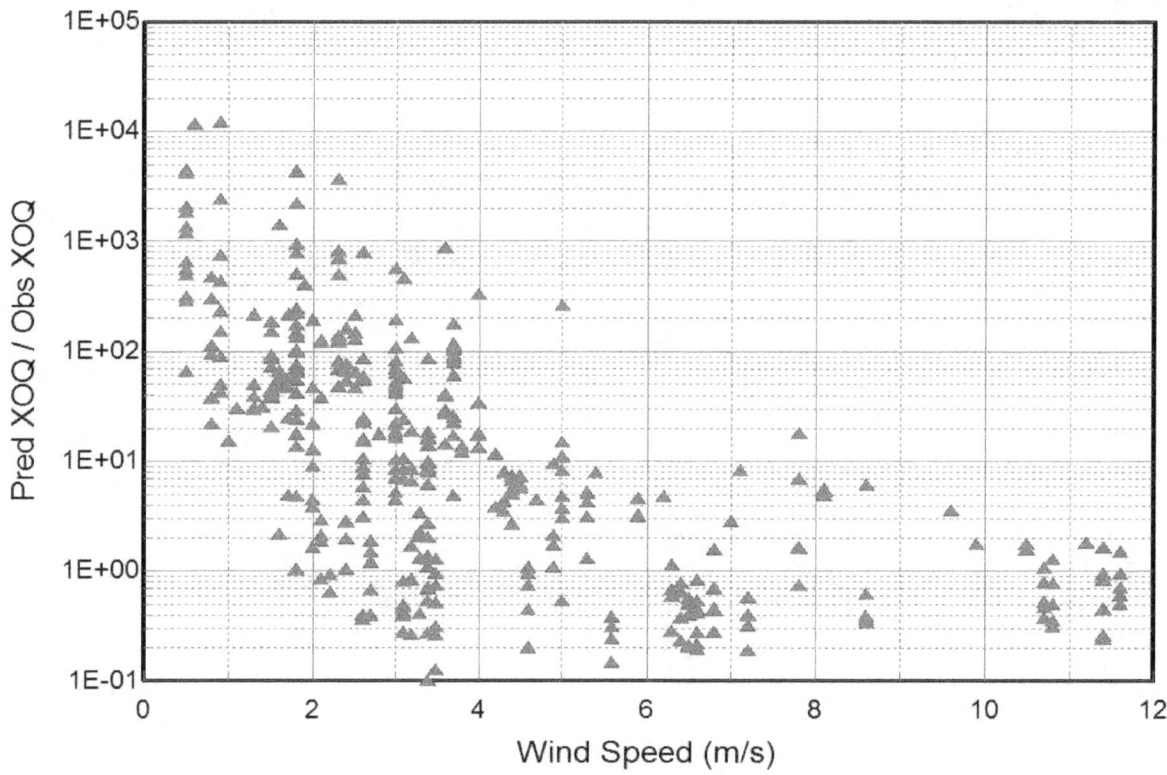

Figure 4.1 Ratios of predicted concentrations in wakes by a model without wake correction to observed concentrations as a function of wind speed.

To account for the underestimation of dispersion at low wind speeds, enhanced dispersion parameters were developed. The enhanced dispersion parameters, Σ_y and Σ_z, are defined as

$$\Sigma_y = \left(\sigma_y^2 + \Delta \sigma_y^2 \right)^{1/2} \tag{4.13}$$

$$\Sigma_z = \left(\sigma_z^2 + \Delta \sigma_z^2 \right)^{1/2} \tag{4.14}$$

and where the enhancement terms, $\Delta\sigma_y^2$ and $\Delta\sigma_z^2$, have the form

$$\Delta \sigma^2(t) = A(1 - (1 + t/T) \exp(-t/T). \tag{4.15}$$

where

$t =$ transport time (distance divided by the wind speed in the plume model and time since release in the puff model),

$A_y =$ $0.5T^2$ (T is the horizontal time scale for wake turbulence),

$A_z =$ $0.5T^2/(2+z/L)$,

$T =$ $B^{1/2}/u^*$,

$B =$ projected building area,

$u^* =$ scaling wind speed (friction velocity) calculated from the wind speed, stability, and surface roughness using the wind profile relationships discussed in Section 6.4.2 ,

$z =$ release height (all releases are assumed to be ≥ 10 m), and

$L =$ Monin-Obukov length which is a function of stability. Evaluation of L is described in Section 6.4.1

Near the release point, the enhancement terms are primarily functions of wind speed and distance, they are weakly dependent on stability, and they are independent of building dimensions. The enhancement terms increase with increasing distance from the release point until they reach an asymptotic limit that is a function of the building area. The terms are large for low wind speeds and decrease as the wind speed increases. They are negligible for wind speeds above about 4 m/s.

This behavior is supported by a more recent analysis of dispersion and turbulence data (Ramsdell and Fosmire 1998). The results of that analysis also suggests that the enhanced dispersion noted in the vicinity of buildings at low wind speeds in wake dispersion experiments may be as much an artifact caused by underestimation of dispersion by the basic dispersion algorithms (Eqs. 4.11 and 4.12) as it is due to increased turbulence in the vicinity of buildings. Consequently, use of the building wake correction option is recommended for all releases other than isolated stacks.

4.3.3 Puff Model Dispersion Parameters

In the puff model, the wind direction, wind speed, and stability are permitted to change as a function of time and position. As a result, puffs may follow curved trajectories and even return to the release point, and stability may change along the puff's trajectory. If dispersion parameters were calculated directly, as in the plume model, there could be discontinuities in the dispersion parameters and possible reduction in plume dimensions with increasing time. Neither is consistent with the known characteristics of atmospheric dispersion. Consequently, dispersion parameters for the puff model can not be directly calculated from Eqs. (4.11) through (4.15) as is done in the plume model.

In the puff model, dispersion parameters are calculated in a four step process as the puffs move through the model domain:

1) Calculate virtual distances from the puff center to the release point based on the current puff dimensions and the meteorological conditions at the position of the puff. The virtual distances are imaginary distance which, if used in Eq. (4.11) (x_{vy}) and Eq. (4.12) (x_{vz}) with the meteorological conditions at the puff position, would give the current dispersion parameters;

2) Add the distance to be moved during the next time step to the virtual distances;

3) Calculate the new dispersion parameters using the sums calculated in step 2; and

4) Add the enhancement term, if appropriate.

Separate virtual distances are required for the horizontal and vertical dispersion parameters because the dispersion parameters increase at different rates. The equations for the virtual distances are

$$x_{vi} = \left(\frac{\sigma_i}{a_i} \right)^{1\ 0.9031} \tag{4.16}$$

and

$$x_{vz} = \left[\frac{\sigma_z - c_z(x)}{a_z(x)} \right]^{1/b_z(x)} \tag{4.17}$$

These equations are just Eqs. (4.11) and (4.12) solved for x rather than σ.

4.3.4 Plume Model Dispersion Parameters

The computational algorithms used in the plume model calculate dispersion parameters directly from Eqs. (4.11) through (4.15) as the parameters are needed. Equations (4.11) and (4.12) are used for plumes released from isolated stacks and for other releases if the building wake correction is not selected. If the building wake options is selected, all five equations are used.

4.3.5 Calm Wind Dispersion Parameters for the Puff Model

The dispersion parameters used during windy conditions increase as the distance traveled increases. During calm or near calm conditions, the distance traveled stops increasing, or increases very slowly. In reality, atmospheric dispersion does not stop when the wind is calm.

The RASCAL 3.0.5 puff model switches from distance-based dispersion parameters to time- based parameters during low wind speed (below 0.5 m/s) conditions to account for continued dispersion. The horizontal dispersion parameter is calculated for each time step is calculated as the parameter for the previous step plus an increment that is only a function of the time step. It is

$$\sigma_y(t + \Delta t) = \sigma_y(t) + \frac{700}{3600} \Delta t \tag{4.18}$$

where

  σ_y = dispersion parameters (m) and,
  Δt = time step (s).

4-12

Clearly then, the 700/3600 (~0.2) is the horizontal rate of growth of the plume (m/s). The horizontal rate of growth of the plume is independent of stability at low wind speeds, just as the exponent of x in Eq. 4.12 is independent of stability.

The growth of the vertical dispersion parameter should be, and is, a function of stability even in low wind speed conditions. Vertical dispersion parameters during low wind speed conditions are calculated by first estimating a fictitious puff movement, Δx, from

$$\Delta x \approx 0.4 \Delta t \qquad (4.19)$$

This fictitious distance is then used, along with the current parameter value, in calculating the dispersion parameter for the next step from Eqs 4.12 and 4.17.

4.3.6 Calm Wind Dispersion Parameters for the Plume Model

Unlike the Gaussian puff model in RASCAL 3.0.5, the straight-line Gaussian plume model does not do well in low wind speeds, and becomes undefined if the wind speed becomes zero. Consequently, RASCAL 3.0.5 changes models when the wind speed falls below 0.5 m/s. The low wind speed model is given in Eq. 4.10. That model requires estimates of horizontal and vertical turbulence velocities (m/s) rather than normal dispersion parameters (m). For the wind speeds of interest, the plume model uses default turbulence velocities $\sigma_u = \sigma_v = \sigma_w = 0.13$ m/s, which are reasonable values for a wind speed less than 1 m/s.

4.4 Mixing Layer Thickness

The mixing layer thickness is included in all dispersion calculations. The thickness is passed to the atmospheric dispersion programs in the meteorological data files that are created by the meteorological data processing program. The meteorological data processing program has three options for determining the mixing layer thickness for each station. The thickness may be estimated from meteorological data and surface roughness; it may be estimated from climatological data; or it may be entered directly. See Section 6.4.3 for details related to estimation of mixing layer thickness.

4.5 Stack Plume Rise

RASCAL 3.0.5 estimates both transition and final plume rise for stacks using Briggs' equations (1969, 1975, and 1984). The transition rise equations are used near the stack until the plume rise exceeds the final plume rise. In general, the numerical constants in the equations are dimensional, and the appropriate metric (SI) units should be assumed.

4.5.1 Final Plume Rise

Plume rise is caused by vertical momentum of the exhaust gases in a stack and buoyancy due to differences in density between the exhaust gases and the atmosphere. Equations for final plume rise for both momentum- and buoyancy-dominated plumes are included in the models. In general, one factor or the other will dominate. For a given set of stack and atmospheric conditions, the temperature difference between the stack effluent and the atmosphere determines which factor is dominant. The initial step in the plume-rise calculation is to determine the critical temperature difference given the stack effluent and atmospheric conditions. If the difference in temperature is less than the critical temperature, plume rise is calculated using momentum-rise equations; otherwise, it is calculated using buoyancy-rise equations. Plume rise is corrected for stack downwash, when the stack exit velocity is less than five times the wind speed.

4.5.1.1 Unstable and Neutral Atmospheric Conditions

In unstable and neutral atmospheric conditions, plume rise is dominated by momentum if the atmospheric temperature is greater than the effluent. When plume rise is dominated by momentum, the plume rise is estimated using Petersen and Lavdas' Eq. A-9 (1986)

$$\Delta h = 6 r_s w_p u(h_s)^{-1} , \qquad (4.20)$$

where

Δh_f = final plume rise estimate (m),
r_s = stack radius (m),
w_p = effluent initial vertical velocity (m/s),
$u(h_s)$ = stack height wind speed (m/s).

If the ambient air temperature is less than the effluent temperature, a critical temperature is calculated from either Petersen and Lavdas' Eq. A-3

$$\Delta t_c = 0.0187 w_p^{1/3} T_p r_s^{-2/3} \qquad (4.21)$$

or Petersen and Lavdas' Eq. A-4

$$\Delta t = 0.00456 \left(\frac{w_p^2}{r_s} \right)^{1/3} T_p , \qquad (4.22)$$

where

Δt_c = critical temperature difference (K) and
T_p = effluent temperature (K).

The choice between Eq. (4.21) and (4.22) is based on the value of a buoyancy-flux parameter, F_b. The buoyancy-flux parameter is defined by Petersen and Lavdas' Eq. A-2 as

$$F_b = g \left[(T_p - T_a)/T_p \right] w_p r_s^2 , \qquad (4.23)$$

where

g = gravitational acceleration (9.8 m/s^2), and

4-14

T_a = air temperature (°K).

If F_b is less than 55, the critical temperature is given by Eq. (4.21), otherwise it is given by Eq. (4.22). Equation (4.20) is used as long as the temperature difference, $T_p - T_a$, is less than Δt_c.

When the temperature difference is greater than Δt_c, the plume rise is calculated using either Petersen and Lavdas' Eq. A-7

$$\Delta h_f = 21.42 F_b^{3/4} u(h_s)^{-1} \tag{4.24}$$

or Petersen and Lavdas' Eq. A-8

$$\Delta h_f = 38.71 F_b^{0.6} u(h_s)^{-1} \ , \tag{4.25}$$

again depending on the value F_b. Equation (4.24) is used when F_b is less than 55, otherwise Eq. (4.25) is used.

4.5.1.2 Stable Atmospheric Conditions

As in unstable and neutral conditions, plume rise is dominated by momentum if the atmospheric temperature exceeds the effluent temperature. If the effluent temperature is greater than the ambient temperature, a Δt_c is calculated to differentiate between momentum- and buoyancy- dominated plumes. The Δt_c given by Petersen and Lavdas' Eq. A-11 is

$$\Delta t_c = 0.0196 w_p T_a S^{1/2} \ , \tag{4.26}$$

where

    S = stability parameter associated with the frequency of vertical oscillation of air parcels in a stable atmosphere.

The square root of S is known as the Brunt-Väisälä frequency discussed in texts on the atmospheric boundary layer, e.g., Panofsky and Dutton (1984) or Stull (1988). It is defined by

$$S = gT^{-1} \frac{\partial \theta}{\partial z} \ , \tag{4.27}$$

where

    $\partial \theta / \partial z$ = potential temperature lapse rate shown in Petersen and Lavdas' Eq. A-10.

Following Petersen and Lavdas, potential temperature lapse rates of 0.02°K/m and 0.035 °K/m are assumed for Pasquill-Gifford stability classes E and F, respectively. A lapse rate of 0.05°K/m is assumed for G stability class following Ramsdell, Simonen, and Burk (1994).

In stable conditions when the temperature difference is less than Δt_c, plume rise is momentum dominated and a plume rise estimate is made using Eq. (4.20). The plume rise also estimated using Petersen and Lavdas' Eq. A-16 is

$$\Delta h = \frac{1.5}{S^{1.}} \left(\frac{F_o w_o T_a}{\pi u(h_s) T_s} \right)^{1/3} , \tag{4.28}$$

where

F_o = stack flow (m³/s).

The two plume rise estimates are compared, and the smaller value is selected.

When plume rise is determined to be buoyancy dominated in stable conditions, the stack-height wind speed is compared with a critical wind speed to determine whether conditions are calm or windy. The critical wind speed, u_c, is calculated using Petersen and Lavdas' Eq. A-15

$$u_c = 0.2746 F_b^{1/4} S^{1/8} . \tag{4.29}$$

If the wind speed is less than the critical speed, the plume rise calculated using Petersen and Lavdas' Eq. A-14 is

$$\Delta h_f = 4.0 F_b^{1/4} S^{-3/8}. \tag{4.30}$$

Otherwise, the rise calculated using Petersen and Lavdas' Eq. A-13 is

$$\Delta h_f = 2.6 \left(\frac{F_b}{u(h_s) S} \right)^{1/3} . \tag{4.31}$$

4.5.2 Transition Plume Rise

Near the stack, the plume gradually increases in height until the final plume rise is reached. RASCAL 3.0.5 calculates the plume rise in this transition area using two relationships presented by Briggs (1984). For plumes without a significant buoyant flux, the transition rise is calculated by

$$\Delta h = \left\{ \frac{3 F_m x}{[\beta_m u(h_s)]^2} \right\}^{1/3} \tag{4.32}$$

where

x = downwind distance, and
$\beta_m = 0.4 + 1.2/r_s$.

If buoyancy is determined to important based on the temperature difference using the tests described above, the transition rise is calculated by

$$\Delta h = \left\{ \frac{3F_m x}{\left[\beta_b u(h_s) \right]^2} + \frac{3F_m x^2}{\left[2\beta_b^2 u(h_s)^3 \right]} \right\}^{1/3}$$ (4.33)

where

β_b has a value of 0.6.

4.5.3 Effective Release Height

If the isolated stack option is selected in RASCAL 3.0.5, the plume height is an effective release height. The effective release height has three components: stack height, plume rise, and stack down wash. It is calculated as

$$h_e = h_s + \Delta h + \Delta h_d \,,$$ (4.34)

where

h_s = stack height,
Δh = smaller of the transitional and final plume rise, and
Δh_d = stack down wash.

The stack height is entered by the user, and the calculation of plume rise is previously described. Stack down wash occurs when the stack exit velocity is less than or approximately the same magnitude as the stack-height wind speed. It is an aerodynamic effect that reduces the stack height by up to three stack diameters. Down wash is calculated in Petersen and Lavdas' Eq. A-1 as

$$\Delta h_d = 4r \left[\frac{w}{u(h_s)} - 1.5 \right]$$ (4.35)

when the ratio between the exit velocity and stack-height wind speed is less than 1.5. Otherwise, the down wash is set to 0.0.

If the isolated stack option is not selected, the release is treated as if it occurs from the surface of the building. A default release height of 10 meters is assumed unless another, greater release height is entered.

4.5.4 Effective Release Height for Fires

If radioactive material becomes involved in a fire, it may be reasonable to consider the release as if it were from an elevated source. In that case, the isolated stack option should be selected and an effective release height should be entered in the form.

4-17

The primary cue for estimating the effective release height is smoke associated with the fire. If the smoke is rising near vertically (low wind speeds), the height of the smoke plume where it levels off may be used as the release height. If the smoke plume is more nearly horizontal (windy conditions), then it may be appropriate to model the release in two parts, one with an estimated release height near ground level (perhaps 10 meters) and the other with an estimated release height near the center of the plume where it levels off.

If the release is modeled in parts, the consequences of the release must be estimated by combining the consequences from the parts as in a weighted average. Unless there is evidence to the support other weights, equal weight should be assigned to each part.

4.6 Deposition

RASCAL 3.0.5 calculates deposition for iodine and particles using the dry and wet deposition models used in MESORAD (Scherpelz, et al. 1986; Ramsdell, et al. 1988) and earlier versions of RASCAL. The activity deposited each time step is the product of the total deposition rate and the time-step duration. At any time, the surface contamination (activity/m^2) is the sum of the activity deposited in the current time step plus previously deposited activity corrected for decay.

4.6.1 Dry Deposition

The dry deposition rate is given by

$$\omega'_d = -v_{dd}\chi \quad ,$$

(4.36)

where

ω'_d = deposition rate in (activity/m^2)/s and
v_{dd} = dry deposition velocity.

RASCAL 3.0.5 assumes a dry velocity of 0.003 m/s (0.3 cm/s). This deposition velocity was used because data summarized by Sehmel (1984) indicate that it is a reasonable value for iodine assuming that about one-third of the iodine in the atmosphere is associated with particles, another one-third is in the from of reactive gases (e.g., I$_2$ or HI) and the remaining one-third is in the form of non-reactive gases (e.g., CH$_3$I). A deposition velocity of 0.003 m/s is slightly high for 1 µm particles. Noble gases (e.g., Kr and Xe) do not deposit.

The assumption that about one-third of the iodine in the atmosphere is associated with particles, another one-third is in the from of reactive gases (e.g., I$_2$ or HI) and the remaining one-third is in the form of non reactive gases (e.g., CH$_3$I) applies only to iodine that has escaped from the containment. The assumption is not appropriate inside the containment. Within the containment most iodine would be in a particulate form. However, the particulates are preferentially removed within the containment. Therefore, the iodine that escapes to the atmosphere is likely to have much higher proportions of reactive and non-reactive gases. In addition, iodine changes form readily in the atmosphere.

4.6.2 Wet Deposition

RASCAL 3.0.5 calculates wet deposition of particles and gases using a washout model with washout coefficients that are only a function of precipitation type and a qualitative measure of intensity. In the washout model, the wet deposition rate is

$$\omega'_w = -\lambda_p \int_0^\infty \chi dz \qquad (4.37)$$

where

λ_p = washout coefficient that is a function of precipitation type.

The washout coefficients used in RASCAL 3.0.5 are listed in Table 4.2. These coefficients are based on analyses of experimental data by Engelmann (1968). Hanna, et al. (1982) and Slinn (1984) point out that the washout model is appropriate strictly for monodisperse aerosols and highly reactive gases only.

Table 4.2 RASCAL 3.0.5 Washout Coefficients

Precipitation Type	Washout Coefficient (h^{-1})
light rain	0.79
moderate rain	2.2
heavy rain	4.0
light snow	0.36
moderate snow	1.2
heavy snow	2.3

Source: Engelmann (1968)

4.6.3 Total Surface Deposition

The total deposition rate at any point is the sum of the dry and wet deposition rates at that point. In the plume model, the total deposition rate is a function of position only because the concentration in the plume is a function of position. In the puff model, the total deposition rate is a function of position because the concentration varies in space and meteorological conditions may vary in space. In both models, the total deposition rate is a function of time because both the concentration and precipitation may change with time.

RASCAL 3.0.5 calculates and stores deposition by radionuclide for all radionuclides except noble gases. Noble gases are assumed not to deposit. However, RASCAL 3.0.5 does calculate the surface contamination from Xe isotopes that result from decay of iodine, assuming that the Xe is trapped within the remaining iodine. The Xe isotopes are not included in ground shine dose calculations. The Cs daughters of Xe are included in the groundshine dose calculations. In general, the contribution of these daughters to the total dose is small.

4.7 Depletion Due to Deposition

Both atmospheric models in RASCAL 3.0.5 calculate depletion caused by wet deposition. Only the puff model accounts for depletion from dry deposition. The plume model does not account for depletion from dry deposition because the calculation is relatively time consuming and dry deposition generally does not result in significant depletion of the plume near the source.

In the puff model, the activity in the puffs (Q) is updated every 5 min. The activity removed from each puff is determined by integrating the total deposition rate under the puff in space and time. In the plume model, the fraction of activity remaining in the plume at each distance during periods of precipitation is estimated assuming an exponential decrease in activity with travel time, i.e., $Q'(x) = Q'_o \exp(-\lambda_p x/u)$.

4.8 Radiological Decay and Ingrowth

The atmospheric dispersion and transport models in RASCAL 3.0.5 calculate radiological decay and ingrowth at 5-minute intervals for both airborne and deposited radionuclides. The decay schemes include as many as four progeny and account for branched decay chains. The 5-minute decay and ingrowth calculations continue to the end of the calculation period specified by the user.

The 5-minute intervals for decay and ingrowth provides reasonably smooth changes in activity for long-lived radionuclides, for example, radionuclides with half-lives of 30 minutes or longer. However, for short-lived radionuclides, such as rubidium-88 ($t_{1/2} = 17.8$ min) or, barium-137m ($t_{1/2} = 2.5$ min), there is a significant decrease in activity present each decay interval. For example, barium-137m activity decreases by about a factor of four every 5 minutes. If short-lived isotopes provide a significant contribution to doses, the step changes in activity may be very evident in the doses and dose rates. Should this artifact of the calculational procedure be observed, the doses and dose rates calculated just after decay calculations are most nearly correct, and doses and dose rates calculated just prior to decay calculations are over estimates.

RASCAL 3.0.5 groundshine dose calculations extend until 96 hours have elapsed from the beginning of the model run specified by the user. Decay and ingrowth calculations during this extended period are made at 1-hour intervals.

4.9 Dose Calculations

RASCAL 3.0.5 calculated doses from inhalation, from groundshine, and from cloudshine.

4.9.1 Inhalation Doses

Committed effective dose equivalents (CEDE) and organ committed dose equivalents (CDE) are calculated for 15-minute periods. These dose equivalents are the sums over all radionuclides of products of the exposure to the radionuclide during the 15-minute period, a radionuclide and organ specific dose factor, and the breathing rate. The general expression for the dose equivalents is

$$D_{15} = \sum \left[v_b DF_n \int_0^{15} \chi_n(t)dt \right] ,$$ (4.38)

where

D_{15} = effective or organ committed dose equivalent due to inhalation during a 15-minute period,
v_b = breathing rate,
DF_n = committed or organ specific dose factor for radionuclide n,
χ_n = radionuclide n concentration,
t = time.

Inhalation dose factors used in RASCAL 3.0.5 are from *Federal Guidance Report No. 11* (Eckerman, Wolbarst, and Richardson 1988). RASCAL 3.0.5 uses a breathing rate of 3.33×10^{-4} m^3/s (20 l/min).

At the end of each 15-minute period, the committed dose equivalents at each receptor node are written to puff and plume model output files. They are then set to zero prior to beginning model calculations for the next 15-minute period.

4.9.2 Groundshine Doses

The puff and plume models calculate groundshine dose equivalents as the sum over all radionuclides of product of the surface contamination by the radionuclide and a radionuclide-specific dose factor. The general expression for the groundshine dose equivalent is
where

$$D_{gs} = \sum_n \left[DF_n \int_t^{t+15} \omega_n(t)dt \right] ,$$ (4.39)

D_{gs} = dose equivalent from groundshine during the period,
DF_n = radionuclide n specific ground-shine dose factor,
ω_n = radionuclide n surface concentration, and
t = time.

The integration in Eq. (4.39) is from t to t+15 minutes. These integration times are used because the surface concentration at any time is cumulative from the beginning of the event and is not set to zero at the beginning of the period. After the surface is contaminated, groundshine doses can be incurred, even if airborne material is not present.

4.9.3 Cloudshine Doses

RASCAL calculates cloudshine doses using four models. The models include a semi-infinite cloud model and a finite-puff model originally developed for MESORAD (Scherpelz, et al. 1986). The first of these models assumes that activity is uniformly distributed through a large volume, and the second assumes that activity is concentrated in a finite number of points distributed through a volume to represent the actual activity distribution. The use of the semi-infinite cloud model is usually inappropriate in the immediate vicinity of the release point, and application of the MESORAD finite-plume model to cloudshine dose estimates near the release point did not prove satisfactory. Consequently, two additional cloudshine dose model have been developed and are used in RASCAL 3.0.5.

The new cloudshine models are based on line sources and plane sources and are analogous to the point-source model. These models are used, along with the point-source model, until plumes and puffs grow to sufficient size that the assumptions associated with the semi-infinite cloud model are met. The finite-plume, cloudshine models in RASCAL 3.0.5 make use of precalculated dose rate vs distance curves. These curves are provided in the radionuclide database for each radionuclide for a 1 Ci (0.01 Sv) point source, and a 1 Ci/m (0.01 Sv/m) infinite-line source. The remainder of this section describes the cloudshine models.

4.9.3.1 Puff Model Cloudshine Dose Calculations

There are three stages in the cloudshine dose calculations. Near the source where puff dimensions are small compared to the mean path length of photons, RASCAL uses a point-kernel dose model. When the puff radius becomes sufficient (σ_y=400 m), cloud-shine dose rates beneath the centerline of the plume are calculated using an infinite-slab model. The dose rate at ground level is calculated assuming that the activity in the plume is equally divided among ten horizontal slabs with slab heights determined by the release height and vertical dispersion coefficients. The change in dose rate with distance from slabs is due only to buildup and absorption of photons; the change in dose rate across the plume is proportional to the crosswind variation of activity concentration in the slab. This model will be discussed further. Finally, when the vertical dimensions of the plume become sufficient (σ_z=400 m), cloudshine is calculated using a semi-infinite cloud model.

The initial versions of RASCAL used the MESORAD finite-puff, cloudshine model (Scherpelz et al. 1986, Ramsdell et al. 1988). This model first calculates composite characteristics (photon energies, photons per disintegration, etc.) of the gamma radiation from the radionuclide mix in a puff. Next, the model calculates the dose rate vs distance from a point source having the composite characteristics using

$$ D_p'(\rho) = \frac{2.13 \times 10^6}{4\pi\rho^2} \sum_\gamma \left[f_\gamma B_\gamma(\mu_{a\gamma},\rho) e^{-\mu_{a\gamma}\rho} E_\gamma T_\gamma W_\gamma \right] \tag{4.40} $$

where

$D_p'(\rho)$ = dose rate in (rem/h)/Ci,
ρ = distance from point source,
f_γ = fraction of disintegrations producing γs of energy, E_γ,
$B_\gamma(\mu_\gamma,\rho)$ = buildup factor for air,
$\mu_{a\gamma}$ = linear attenuation factor for air,
E_γ = gamma energy,
T_γ = mass energy absorption coefficient for tissue ($\mu_{t\gamma}/\rho_t$),
W_γ = ratio of whole body dose to surface dose.

The constant 2.13×10^6 is a collection of unit conversion constants to give dose rate in (rem/h)/Ci. The components of the constant are described following Eq. 19 in Scherpelz, et al. (1986).

The model then calculates the dose rate at ground level as a function of horizontal distance from the ground-level position of center of the puff. This calculation involves summation over volume elements distributed throughout the puff.

$$D'(r) = \sum_i \sum_j \sum_k D'_p(\rho_{ijk}) M_{ijk} \quad , \tag{4.41}$$

where

$D'(r)$ = dose rate at r,
r = distance from the receptor to the projection of the puff center on the ground,
i,j,k = indices associated with the volume elements,
$D'_p(\rho_{ijk})$ = dose rate at distance ρ from a point source in volume element ijk,
ρ_{ijk} = distance from the center of the volume element ijk to the receptor,
M_{ijk} = fraction of the total puff activity in volume element ijk.

For purposes of the cloudshine calculation, puffs were assumed to be circular cylinders with three layers. The volume elements were defined in 3, 5, or 8 annular rings with either 6 or 16 sectors. The fraction of activity in volume elements varied by annulus and level. Finally, the dose at a receptor for a period is accumulated by summing the product of dose rates and the time step for all puffs for all time steps in the period.

In RASCAL 3.0.5, this process has been modified by eliminating the calculation of composite characteristics from the gamma energies. In its place, the puff model calculates the dose rate vs distance from a point source that has all of the activity in the puff. Thus, Eq. (4.40) is replaced by

$$D'_p(\rho) = \sum_{n=1}^{N} Q_n D'_{pn}(\rho) \tag{4.42}$$

where

N = number of radionuclides,
Q_n = activity of radionuclide n in the puff,
$D'_{pn}(\rho)$ = dose rate at distance ρ from a 1-Ci point source of radionuclide n.

The overall puff geometry remains the same in the puff model as it was in earlier versions of RASCAL. However, the internal geometry has changed. The puff is divided into ten layers with each layer containing one-tenth of the activity. The number of annular rings has been fixed at six, with each ring containing one-sixth of the puff activity, and the number of sectors has been fixed at 12. With these changes, the number of volume elements has been increased, and the fraction of activity in each volume element becomes 1/720 of the total activity. In the puff model, Eq. (4.41) becomes

$$D'(r) = \frac{1}{720} \sum \sum \sum D'_p(\rho_{ijk}) \quad . \tag{4.43}$$

As before, symmetry is used to reduce the computational load.

When the horizontal dispersion parameter reaches 400 m, the puff radius is large enough that the horizontal variations in the cloud-shine dose rate are directly proportional to the horizontal variation in concentration in the puff. At this point, MESORAD changed to a semi-infinite cloud model, and previous versions of RASCAL changed to calculating dose rates beneath the center of the puff using the finite-puff model and dose rates elsewhere using the horizontal variation in concentration. The puff model changes from the point-source based cloud-shine model used in earlier versions of RASCAL to a new plane-source based model. The plane-source model assumes that the puff is a vertical cylinder as is assumed in the point-source model. However, rather than assuming that activity is distributed among volume elements, the activity is assumed to be concentrated on ten horizontal slabs (planes).

To calculate the activity in each slab, the concentration at the center of the puff (y=0) is first integrated vertically from the bottom of the puff to the top. This is similar to the integration done in calculating the depletion from wet deposition

$$\langle \chi \rangle_n = \int_{-\infty}^{\infty} \frac{Q}{(2\pi)^{3/2} \sigma^2 \sigma_z} \exp\left[-0.5\left(\frac{r}{\sigma}\right)^2\right] \exp\left[-0.5\left(\frac{h}{\sigma_z}\right)^2\right] dz = \frac{Q}{2\pi\sigma^2} \exp\left[-0.5\left(\frac{r}{\sigma}\right)^2\right], \qquad (4.44)$$

where

$\langle \chi \rangle_n$ = vertically integrated concentration of radionuclide n at the center of the puff.

This concentration is then divided by the number of slabs (ten) to get the concentration in each slab.

Within the cylinder, the vertical position of the slabs is determined by the effective release height, the mixing-layer thickness, and the vertical-dispersion coefficient. When a Gaussian distribution is partitioned so the area under the curve is divided into ten equal parts and the center of mass of each part is determined, these centers of mass fall at $\pm0.127\sigma$, $\pm0.385\sigma$, $\pm0.674\sigma$, $\pm1.037\sigma$, and $\pm1.645\sigma$. Using this as a basis, the slab heights are initially estimated as $h_e\pm0.127\sigma_z$, $h_e\pm0.385\sigma_z$, $h_e\pm0.674\sigma_z$, $h_e\pm1.037\sigma_z$, and $h_e\pm1.645\sigma_z$. The initial heights may lie below ground level or above the top of the mixing-layer. Any heights that fall outside these bounds are adjusted to account for reflection by the boundaries. Signs of heights that are negative are changed to positive, and heights (h_p) that are above the mixing layer are replaced by 2H-h_p.

The dose rate at ground level from a slab is calculated as

$$D'_n(r,z) = \frac{0.1\sum Q\, DF_{pn}}{2\pi\sigma^2} \exp\left[-0.5\left(\frac{r}{\sigma}\right)^2\right]\left[1 + k\mu z\, |\exp(-\mu z)\right], \qquad (4.45)$$

where

z = height of the slab above the receptor, which is assumed to be at 1 m (m),
DF_{pn} = dose factor radionuclide n for an infinite plane [(rem/s)/(Ci/m²)],
μ = total gamma ray absorption coefficient for air (m⁻¹),
k = ratio of energy in scattered photons to absorbed energy.

The infinite plane dose factor is approximated by

$$DF_{pn} = DF_{sicn}/241.2 \; ,$$ (4.46)

where

DF_{sicn} = semi-infinite cloud-dose factor $[(rem/s)/(Ci/m^3)]$ and
241.2 = constant with units of meters evaluated by comparing dose rates calculated by Eq. (4.45) with semi-infinite cloud-dose rates in plumes for which the semi-infinite cloud model is appropriate.

Semi-infinite dose factors contained in *Federal Guidance Report No. 12* (Eckerman and Ryman 1993) are used to estimate the infinite-plane dose factors. These calculations were carried out for 30 radionuclides that are typically released in reactor accidents involving fuel damage. The standard deviation of the estimates of the constant value was 0.04.

In Eq. (4.45), the term $(1+k\mu z)$ represents the buildup factor due to scattered photons, and $exp(-\mu z)$ represents the absorption of energy by the air. These terms are discussed by Healy and Baker (1968) and Healy (1984). In RASCAL 3.0.5, μ and k are assumed to be constants with values appropriate for ~0.7 MeV photons (μ=0.01, k=1.4) based on Figure 16.4 of Healy (1984).

Ultimately, the dose rate at a receptor is

$$D'(r) = \frac{0.1 \sum_n Q_n DF_{pn}}{2\pi\sigma_z^2} exp\left[-0.5\left(\frac{r}{\sigma_r}\right)^2\right] \sum_{-1}^{10} (1 + k\mu z) \, exp\,(-\mu z) \;,$$ (4.47)

where the summation is over all slabs.

When the vertical extent of the puff is sufficient for the semi-infinite cloud model to be appropriate (σ_z > 400 m, or a uniformly mixed plume with a vertical depth >600 m), the cloudshine dose rate is calculated using the semi-infinite cloud model

$$D'(r) = [\chi(r)/Q]\sum_n Q_n DF_{sicn} \;,$$ (4.48)

where

$[\chi(r)/Q]$ is calculated using Eq. (4.2) or a variation thereof, as appropriate.

The semi-infinite dose factors from *Federal Guidance Report No. 12* (Eckerman and Ryman 1993) are used in this calculation.

4.9.3.2 Plume Model Cloudshine Dose Calculations

In the plume model, near the source, the plume is divided into a large number of equal-strength line sources spaced to properly represent the distribution of activity in the plume. The dose rates from these line sources are used to calculate the ground-level dose rate as a function of horizontal distance from the plume axis. This relationship is then used to calculate dose rates and 15-minute doses at receptor locations. When the width of the plume is sufficient (σ_y = 400 m), the plume model switches from the line-source model to an infinite-slab model. Finally, when the vertical dimensions of the plume are sufficient to make the semi-infinite cloud model appropriate (σ_z = 400 meters or a uniformly mixed plume with a thickness of 600 meters), the plume model switches to a semi-infinite cloud model.

Dose rates from line sources are calculated using

$$D'_l(\rho) = \sum_{n=1}^{N} Q'_{ln} D'_n(\rho) , \qquad (4.49)$$

where

$D'_l(\rho)$ = dose rate (rem/s) at a distance ρ from an infinite line source of Q'_{ln} (Ci/m),
Q'_{ln} = line-source strength (Ci/m), $Q'_{ln} = Q'_n / u$ where Q'_n is in Ci/s,
$D'_n(\rho)$ = line-source dose rate factor [(rem/s)/(Ci/m)] for radionuclide n.

Equation (4.49) is analogous to Eq. (4.42) with changes in the definitions of source term and dose factors.

The line-source dose rates are combined to get the plume dose rate by summing over all line sources, just as the point-source dose rates were combined to get a puff dose rate. The number of line sources used is determined by the horizontal dispersion parameter, σ_y. If σ_y > 200 m, 100 lines (10×10) are used to describe the concentration distribution in the plume. Otherwise, the concentration distribution is described by 36 lines (6×6). In either case, the lines are spaced horizontally and vertically such that each line represents the same fraction of the total activity in the plume. The plume dose rate is given by

$$D'(y) = \frac{C_R}{N_l} \sum_i \sum_j D'(y_{ij}) , \qquad (4.50)$$

where

$D'(y)$ = plume dose rate ,
y = distance from the ground-level projection of the center of the plume,
C_R = finite line correction factor,
N_l = number of line sources (36 or 100),
i,j = line source indices,
$D'_l(y_{ij})$ = infinite line, line-source dose rate.

A finite-line source correction factor is included in Eq. (4.50) to account for the fact that the plume does not extend upwind of the release point. A correction factor could be calculated by numerical integration of a rather complex equation. However, an approximate correction factor of adequate accuracy for emergency response dose calculations can be estimated using

$$C_R = 0.5 \left[1 + \frac{x}{\left(R^2 + h_e^2 \right)^{1/2}} \right] ,$$ (4.51)

where

x = downwind distance (m) to a point beneath the plume centerline at the intersection of the plume centerline and a perpendicular line passing through the receptor,
R = distance (m) from the release point to the receptor,
h_e = effective release height.

In the case of a ground-level release and a receptor on the plume centerline, the correction factor will be one. However, generally the correction factor is less than one. For a 0.7 MeV photon, the correction factor given by Eq. (4.51) corresponds to a receptor at a position approximately 100 m off of the plume centerline. Doses will be slightly over estimated for receptors that are closer than 100 m and slightly under estimated for receptors that are farther than 100 m from the centerline.

When the horizontal dispersion parameter exceeds 400 m, the plume model shifts from a line-source based, finite-plume model to an infinite-plane model. The infinite-plane model is similar to the model used by in the puff model. The differences between the two models are associated with the calculation of concentrations, not with the cloudshine calculation. Thus, Eq. (4.47) for the plume model becomes

$$D'(r) = \frac{0.1 \sum Q'_n DF_{xn}}{2 \pi \sigma_z u} \exp \left[-0.5 \left(\frac{r}{\sigma_z} \right)^2 \right] \sum_{-1}^{10} \left(1 + k \mu z \right) \exp \left(-\mu z \right)$$ (4.52)

Finally, when the vertical extent of the plume is sufficient (σ_z = 400 m, or a uniformly mixed plume with a 600 m vertical extent), the plume model shifts to a semi-infinite plume, cloudshine model. Equation (4.48) for the plume model becomes

$$D'(r) = [\chi(r)/Q'] \sum_n Q'_n DF_{sxn} ,$$ (4.53)

$\chi(r)/Q'$ is calculated using Eq. (4.5) or a variation thereof, as appropriate.

4.9.4 Open- and Closed-Window Dose Rates

The open- and closed-window dose rates, which are reported in mrad/h, are intended for use in comparisons with field radiation measurements. The open-window dose rate has four components. These components are gamma and beta radiation from airborne activity and gamma and beta radiation from surface contamination. The closed-window dose rates have only the gamma radiation components.

Beta dose rates are calculated using the semi-infinite cloud model with appropriate dose factors. Similarly, the gamma and beta groundshine dose rates are calculated using the models used to calculate external doses to the body, with dose factors for air in place of dose factors for tissue. Dose factors used in these calculations were extracted from the original data used in producing *Federal Guidance Report No. 12* by its authors, and are included in the RASCAL 3.0.5 radionuclide database.

The gamma dose rates for airborne activity in finite plumes are calculated using the same algorithms used in the puff model and the plume model for finite-plume, cloudshine doses with adjustment for the difference in energy absorption coefficients of tissue and air. The adjustment is made to the dose rates for each isotope by multiplying the dose rate for tissue by the ratio of the semi-infinite cloud dose factor for air and the semi-infinite cloud dose factor for tissue. Thus, the gamma dose rate for an isotope is

$$D'_{air} = D' \frac{DF_{air}}{DF_{tissue}} \quad , \qquad (4.54)$$

where D' is the cloudshine dose rate calculated for tissue. Semi-infinite cloud-dose factors for tissue from *Federal Guidance Report No. 12* (Eckerman and Ryman 1993) are used in this calculation along with semi-infinite cloud-dose factors for air calculated by A. L. Sjoreen using the methods listed in (Scherpelz 1986). The semi-infinite dose factors for air are included in the RASCAL 3.0.5 radionuclide database.

The open- and closed-window dose rates are larger than the sum of the cloudshine and groundshine doses because the dose factors for air are larger than those for tissue. Typically, the cloudshine gamma dose rate for air is about a factor 1.4 larger than the dose rate for tissue, and the groundshine gamma dose rate for air is about a factor 1.3 larger than the dose rate for tissue. These ratios may be significantly larger greater than 2) if radionuclides with low energy gamma emissions (<0.1 MeV) contribute significantly to the dose rates.

4.9.5 Total Effective Dose Equivalent

The early phase (plume phase) total effective dose equivalent (TEDE) that RASCAL 3.0.5 calculates is the sum of the external gamma dose (cloudshine) from the plume, the committed effective dose equivalent (CEDE), and the external dose over a four-day period from radionuclides deposited on the ground (4-day groundshine dose). This TEDE is calculated assuming that no protective actions such as evacuation or sheltering are taken. Thus, the calculations assume that people are outdoors during plume passage and will remain outdoors exposed to ground shine from deposited radionuclides for four days after the radionuclides have been deposited.

Thus, the early phase TEDE that RASCAL 3.0.5 calculates is larger than the TEDE that would be expected for people who took protective actions or who continued their normal activities (spending much time indoors).

The reason that RASCAL 3.0.5 calculates dose assuming that no actions to reduce dose are taken is to determine if doses without any protective actions would exceed the EPA protective action guides. The need for protective actions is based on the TEDE that would be received if no protective actions of any type were taken, even actions such as simply spending some time indoors.

The RASCAL 3.0.5 dose estimates should not be used as an estimate of the TEDE that would be received by people who did not intentionally take protective actions because even performing normal everyday activities will reduce doses to below those estimated by RASCAL 3.0.5.

RASCAL 3.0.5 can provide a more realistic estimate of the doses that people would actually receive, but that requires some effort. To account for evacuation, the end of calculation time can be set to the time at which people evacuate. The TEDE then is the sum of the inhalation dose, cloudshine dose, and *period* ground shine dose. If sheltering prior to evacuation is to be taken into account, each dose component must be reduced by an appropriate reduction factor for sheltering before the three dose components are summed.

4.9.6 Total Acute Bone Dose Equivalent

Total acute bone dose equivalent is another multiple pathway dose calculated in RASCAL 3.0.5. The total acute bone dose equivalent is the sum of the cloudshine dose, the ground shine dose, and a 4-day inhalation dose to bone red marrow.

4.10 References

Athey, G. F., et al. 1993. *RASCAL Version 2.0 User's Guide.* Vol. 1., Rev. 1., NUREG/CR-5247, U.S. Nuclear Regulatory Commission. Washington, D.C.

Bander, T. J. 1982. *PAVAN: An Atmospheric Dispersion Program for Evaluating Design Basis Accidental Releases for Radioactive Materials from Nuclear Power Stations.* NUREG/CR-2858, U.S. Nuclear Regulatory Commission. Washington, D.C.

Briggs, G. A. 1969. *Plume Rise*, TID-25075, U.S. Atomic Energy Commission. Washington, D.C.

Briggs, G. A. 1975. "Plume Rise Predictions," *Lectures on Air Pollution and Environmental Impact Analyses.* American Meteorological Society, Boston, Mass.

Briggs, G. A. 1984. "Plume Rise and Buoyancy Effects." *Atmospheric Science and Power Production.* Ed. D. Randerson, DOE/TIC-27601, U. S. Department of Energy.

Csanady, G. T. 1973. *Turbulent Diffusion in the Environment.* D. Reidel, Boston, Mass.

Draxler, R. R. 1984. "Diffusion and Transport Experiments." *Atmospheric Science and Power Production.* Ed. D. Randerson, DOE/TIC-27601, U. S. Department of Energy. Washington, D.C.

Eckerman, K. F., A. B. Wolbarst, and A. B. C. Richardson. 1988. *Federal Guidance Report No. 11. Limiting Values of Radionuclide Intake and Air Concentration and Dose Conversion Factors for Inhalation, Submersion, and Ingestion.* U. S. Environmental Protection Agency. Washington, D.C.

Eckerman, K. F., and J. C. Ryman. 1993. *Federal Guidance Report No. 12. External Exposure to Radionuclides in Air, Water, and Soil.* U. S. Environmental Protection Agency. Washington, D.C.

Eimutis, E. C., and M. G. Konicek. 1972. "Derivations of Continuous Functions for the Lateral and Vertical Atmospheric Dispersion Coefficients." *Atmospheric Environment,* **6**:859-63.

Englemann, R. J. 1968. "Calculation of Precipitation Scavenging." *Meteorology and Atomic Energy 1968.* Ed. D. Slade, TID-24190, U.S. Atomic Energy Agency.

Frenkiel, F. N. 1953. "Turbulent Diffusion: Mean Concentration in a Flow Field of Homogeneous Turbulence." *Advances in Applied Mechanics,* **3**:61–107.

Gifford, F. A. 1976. "Turbulent Diffusion-Typing Schemes: A Review." *Nuclear Safety,* **17** No. 1: 68–86.

Hanna, S. R., G. A. Briggs, and R. P. Hosker. 1982. *Handbook on Atmospheric Diffusion.* DOE/TIC-11223, U.S. Department of Energy. Washington, D.C.

Healy, J. W., and R. E. Baker. 1968. "Radioactive Cloud-Dose Calculations." *Meteorology and Atomic Energy 1968.* Ed. D. Slade, TID-24190, U.S. Atomic Energy Agency. Washington, D.C.

Healy, J. W. 1984. "Radioactive Cloud-Dose Calculations." *Atmospheric Science and Power Production.* Ed. D. Randerson, DOE/TIC-27601, U. S. Department of Energy. Washington, D.C.

Kao, S. K. 1984. "Theories of Atmospheric Transport and Diffusion." *Atmospheric Science and Power Production.* Ed. D. Randerson, DOE/TIC-27601, U.S. Department of Energy.

Martin, D. O., and J. A. Tikvart. 1968. "A General Atmospheric Diffusion Model for Estimating the Effects on Air Quality of One or More Source." *61ˢᵗ Annual Meeting of the Air Pollution Control Association for NAPCA,* St. Paul, Minnesota.

Panofsky, H. A., and J. A. Dutton. 1984. *Atmospheric Turbulence.* J. Wiley & Sons, New York.

Petersen, W. B., and L. G. Lavdas. 1986. *INPUFF 2.0 - A Multiple Source Gaussian Puff Dispersion Algorithm User's Guide.* EPA/600/8-86/024, Atmospheric Sciences Research Laboratory, U.S. Environmental Protection Agency, Research Triangle Park, N.C.

Press, W. H., et al. 1986. *Numerical Recipes: the Art of Scientific Computing.* Cambridge University Press, Cambridge, United Kingdom.

Ramsdell, Jr., J. V., et al. 1988. *The MESORAD Dose Assessment Model, Volume 2: Computer Code.* Vol. 2., NUREG/CR-4000, U.S. Nuclear Regulatory Commission. Washington, D.C.

Ramsdell, Jr., J. V. 1990. "Diffusion in Building Wakes for Ground-Level Releases." *Atmospheric Environment,* **24B**:377–88.

Ramsdell, Jr, J. V., C. A. Simonen, and K. W. Burk. 1994. *Regional Atmospheric Transport Code for Hanford Emission Tracking (RATCHET).* PNWD-2224 HEDR, Battelle, Pacific Northwest Laboratories, Richland, Wash.

Ramsdell, Jr., J. V., and C. J. Fosmire. 1998. "Estimating Concentrations in Plumes Released in the Vicinity of Buildings: Model Development." *Atmospheric Environment,* **32**:1663–17.

Randerson, D. 1984. *Atmospheric Science and Power Production.* DOE/TIC-27601, U. S. Department of Energy.

Sagendorf, J. F., J. T. Goll, and W. F. Sandusky. 1982. *XOQDOQ: Computer Program for the Meteorological Evaluation of Routine Effluent Releases at Nuclear Power Stations.* NUREG/CR-4380, U.S. Nuclear Regulatory Commission. Washington, D.C.

Scherpelz, R. I., et al. 1986. *The Mesorad Dose Assessment Model.* Vol. 1., NUREG/CR-4000, U.S. Nuclear Regulatory Commission. Washington, D.C.

Sehmel, G. A. 1984. "Deposition and Resuspension." *Atmospheric Science and Power Production.* Ed. D. Randerson, DOE/TIC-27601, U. S. Department of Energy. Washington, D.C.

Seinfeld, J. H. 1986. *Atmospheric Chemistry and Physics of Air Pollution.* John Wiley & Sons, New York.

Slade, D. H. 1968. *Meteorology and Atomic Energy 1968.* TID-24190, U.S. Atomic Energy Agency. Washington, D.C.

Slinn, W. G. N. 1984. "Precipitation Scavenging." *Atmospheric Science and Power Production.* Ed. D. Randerson, DOE/TIC-27601, U. S. Department of Energy. Washington, D.C.

Stull, R. B. 1988. *An Introduction to Boundary Layer Meteorology.* Kluwer Academic Publishers, Dordrecht, Netherlands.

Tadmor, J., and Y. Gur. 1969. "Analytical Expressions for Vertical and Lateral Dispersion Coefficients in Atmospheric Diffusion." *Atmospheric Environment*, **3**:688-98.

5 UF₆ Transport and Diffusion Model

RASCAL 3.0.5 contains a special version of the plume model (see Chapter 4) modified to treat releases of UF_6. The modifications include the introduction of a dense gas model to treat the gravitationally driven spread of UF_6 releases, a chemical/thermodynamic model to treat the reaction of UF_6 with water (both liquid and vapor) in the atmosphere, and a plume rise model to treat the vertical displacement of HF/UO_2F_2 plumes when their densities become less than the density of air.

The dense gas and chemical/thermodynamic models are implemented in two control volumes, one for UF_6 and a second for HF and UO_2F_2. "Control volumes" as used in thermodynamics, are volumes in which mass, energy, moisture, etc. are evaluated taking into account the quantities moving into and out of the volume. The control volumes move downwind at the speed of the wind 1 meter above ground level. The size of the control volumes are initially defined by the release rates of UF_6, HF, and UO_2F_2. As the control

$$UF_6 + 2H_2O \rightarrow UO_2F_2 + 4HF + heat \qquad (5.1)$$

volumes move downwind, the volume of UF_6 is deformed by gravitational settling, and air and water vapor are mixed into the UF_6 volume. The chemical reaction is assumed to occur instantaneously as the mixing takes place. The result of the reaction is a decrease of mass and volume of the control volume containing UF_6 and increase in the mass and volume in the HF/UO_2F_2 control volume. The temperatures of these two control volumes are assumed to be identical and are determined from the initial temperature of the released material, the air temperature, and the heat reaction of UF_6 and water.

The output of the dense gas and chemical/thermodynamic model calculations is used as input to atmospheric dispersion and deposition calculations. This input is a function of the distance from the release point to the point at which all the UF_6 has been converted to HF and UO_2F_2. After the UF_6 is gone, the HF and UO_2F_2 source terms continue to decrease with distance to account for deposition as described in Chapter 4.

5.1 UF₆ Model Assumptions and Equations

The following assumptions were made in the development of the UF_6 model.

1. The UF_6 plume is released at or near ground level. (Elevated releases are not modeled.)

2. An initial UF_6 control volume is defined by the UF_6 release rate and density.

3. The initial cross section of the UF_6 control volume is square with .

$$A_{UF6} = \frac{Q'_{UF6}}{\rho_{UF6} u} , \qquad (5.2)$$

where

A_{UF6} = cross-sectional area (m^2),
Q'_{UF6} = UF_6 release rate (g/s),
ρ_{UF6} = UF_6 density (g/m^3),
u = wind speed at 1 m (m/s),

If the release includes HF and UO_2F_2 in addition to UF_6,. the area of the initial control volume is given by

$$A_{cv} = \frac{V'_{UF6} + V'_{HF} + V'_{air}}{u} \qquad (5.3)$$

where

V'_{UF6} = the release rate of UF_6 (m^3/s)
V'_{HF} = the release rate of HF (m^3/s)
V'_{air} = the volume flow of air that would be needed to generate the HF flow from a reaction of the air with UF_6 (m^3/s)
u = wind speed at 1 m (m/s).

4. There is no diffusion of the UF_6 plume.

5. Deformation of the UF_6 control volume is determined by gravitational slumping of the UF_6.

6. The rate of change of the UF_6 control volume width is given by

$$\frac{dw_{UF6}}{dt} = k\left[g\frac{(\rho_{UF6} - \rho_{air})}{\rho_{UF6}}H_{UF6}\right], \qquad (5.4)$$

where

w_{UF6} = UF_6 control volume width (m),
t = time (s),
k = a slumping constant (dimensionless),
g = gravitational constant (m/s^2),
ρ_{air} = density of air (g/m^3),
H_{UF6} = thickness of the control volume (m).

7. The slumping constant has a theoretical value of 1.4 ($2^{1/2}$) (Eidsvik 1980) but may be given a lower value to account for surface resistance or to tune the model. A value of 1.3 is used as default in the current version of the UF_6 model in RASCAL.

8. Air is entrained into the UF_6 control volume only through the top. Entrainment through the sides is negligible because after only a few seconds the area of the top of the volume is much larger than the area of the sides.

9. The rate of entrainment of air into the UF_6 is given by

$$\frac{dV_{air}}{dt} = u_e w_{UF6} u \ ,$$

(5.5)

where

V_{air} = air entrainment rate (m³/s),
u_e = an entrainment velocity (m/s).

10. The entrainment velocity u_e is given by

$$u_e = \frac{\rho_{air} u_*^3}{(\rho_{UF6} - \rho_{air}) g h_{UF6}} \ ,$$

(5.6)

where u_* is a scaling velocity (m/s) associated with atmospheric turbulence.

11. The water available for reaction with UF_6 is determined by a combination of the water vapor in the entrained air and precipitation entering the UF_6 control volume.

12. The water available for reaction is given by

$$m_{H2O} = \rho_{H2Ov} V_{air} + p_r w_{UF6} u \Delta t \rho_{H2Ol} \ ,$$

(5.7)

where

Δt = the duration of the time step (s)
m_{H2O} = the rate at which water becomes available for reaction (g/s),
ρ_{H2Ov} = density of water vapor in the ambient air (g/m³),
p_r = precipitation rate (m/s),
ρ_{H2Ol} = density of liquid water (g/m³).

13. The reaction between UF_6 and water is assumed to occur at the top of the UF_6 control volume. The volume of UF_6 involved in the reaction is subtracted from the UF_6 control volume, and the masses of air, HF, and UO_2F_2 are added to the HF/UO_2F_2 control volume. The volume of the HF/UO_2F_2 control volume is increased by the volumes of the air and HF. The UO_2F_2 formed in the UF_6/H_2O reaction is present as small particles that are assumed to have negligible volume. The temperatures and volumes of the control volumes are adjusted to conserve enthalpy in a constant pressure reaction.

14. Potential heat exchange with the ground and possible reaction of UF_6 with water on the ground surface are assumed to be negligible.

15. The ground is assumed to be a sink for UF_6 that may be deposited on the ground. Any UF_6 condensing in the UF_6 control volume is assumed to deposit on the ground. In addition, 25% of the UO_2F_2 formed in the UF_6/H_2O reaction is assumed to deposit at the time of the reaction, unless the UF_6 is released in a fire. Wet deposition of UF_6 is not modeled because all water entering the UF_6 control volume is assumed to react with UF_6 to produce HF and UO_2F_2.

5.2 Chemical/Thermodynamic Model

The chemical/thermodynamic model in the UF_6 plume model is based on the description contained in NUREG/CR-4360, *Calculational Methods for Analysis of Postulated UF_6 Releases* (Williams 1985). The initial release to the atmosphere may be UF_6 or a mixture of UF_6, HF, and UO_2F_2. However, the chemical/thermodynamic model is invoked only when the release includes UF_6. A release of HF and UO_2F_2 is treated as a release of passive contaminants.

Air, water vapor, and HF are assumed to be ideal gases. A compressibility factor is used to account for the deviation of UF_6 behavior from that of an ideal gas. Although UF_6 cannot exist as a liquid at atmospheric pressures, equations for the density, vapor pressure, and enthalpy of liquid UF_6 are included in the UF_6 plume model because they were included in the computer code published by Williams (1985).

5.2.1 Compressibility Factor

Dewitt (1960) cites work by D. W. Magnuson in presenting the following relationship for a UF_6 compressibility factor

$$Z = \frac{T_r^3}{\left(T_r^3 + 4.892 \times 10^5 \, P\right)} ,$$

(5.8)

where

Z = the compressibility factor, (dimensionless)
T_r = the temperature (°R),
P = the pressure (psia)
4.892 = a constant with the dimensions (°R³/psia)

5.2.2 UF_6 Density

$$\rho_{UF6s} = 330.0 - 0.180 \, T_f \left(\frac{MW}{352}\right) ,$$

(5.9)

The density of UF_6 is given by the following relationships. The relationships for the UF_6 liquid and vapor are based on the work of Dewitt (1960), and the relationship for the density of UF_6 solid was derived by Williams (1985) based on data presented by Dewitt.

The density of solid UF_6 is given by
where

ρ_{UF6s} = the density of the solid UF_6 (lb_m/ft^3),
T_f = the temperature (°F),
MW = the molecular weight of UF_6.

The density of liquid UF_6 is given by

$$\rho_{UF6l} = \left(250.6 - 0.1241T_f + 2.620 \times 10^{-4}T_f^2\right)\left(\frac{MW}{352.0}\right)$$ (5.10)

where
ρ_{UF6l} = the density of the liquid UF_6 (lb_m/ft^3),
T_f = the temperature (°F),
MW = the molecular weight of UF_6.

The density of UF_6 vapor is given by

$$\rho_{UF6v} = \frac{MW \cdot P \cdot Z}{R \cdot T},$$ (5.11)

where

R = universal gas constant, 10.73 (psia -ft^3)/(lb-mol °R).

5.2.3 UF_6 Vapor Pressure

The following relationships, based on the work of Dewitt (1960), describe the vapor pressure of UF_6. The constants in the relationships assume English units for pressure, temperature, and volume.

From 32°F to the triple point of 147.3°F, the vapor pressure of UF_6 in the solid phase is

$$P_{UF6s} = \exp\left[10.44 + 9.642 \times 10^{-3}T_f - \frac{3.90 \times 10^3}{\left(T_f + 298.1\right)}\right],$$ (5.12)

where

P_{UF6s} = vapor pressure (psia) , and
T_f = temperature (°F).

From the triple point (147.3°F) to 240°F, the vapor pressure is given by

$$P_{UF6vl} = \exp\left[12.16 - \frac{4.668 \times 10^3}{\left(T_f + 367.5\right)}\right]$$ (5.13)

And, from 276°F to the critical temperature (446°F) the vapor pressure is given by

5-5

$$P_{UF6vh} = \exp\left[13.76 - \frac{6.976 \times 10^3}{\left(T_f + 5119\right)}\right]$$ (5.14)

Between 240°F and 276°F, the vapor pressure is estimated by a weighted average of P_{uf6vl} and P_{uf6vh}

$$P_{UF6v} = P_{UF6vl}\left(276.0 - T_f\right) + P_{UF6vh}\left[\frac{\left(T_f - 240.0\right)}{36.0}\right]$$ (5.15)

5.2.4 UF$_6$ Enthalpy

Williams (1985) provides the following equations for the enthalpy of UF$_6$ using 25°C (77°F) as a base. The equations are to a large extent based on data of Dewitt (1960).

For solid UF$_6$, the enthalpy is given by

$$H_{UF6s} = 50.446 - 5.70531 \times 10^{-2} T_r + 1.27509 \times 10^{-4} T_r^2 - 9645.63 T_r^{-1},$$ (5.16)

where H_{UF6s} is the enthalpy (Btu/lb$_m$).

For liquid UF$_6$, the enthalpy is given by

$$H_{UF6l} = 30.6133 + 5.10057 \times 10^{-2} T_r + 5.13165 \times 10^{-5} T_r^2$$
$$- 6.139.34 T_r^{-1} + 0.18268\left[\frac{\left(P - P^o\right)}{r_l}\right],$$ (5.17)

where

    H_{UF6l} = the enthalpy,
    P = the atmospheric pressure (psia),
    P^o = the vapor pressure over liquid UF$_6$ (psia),
    ρ_l = the density of the liquid (lb$_m$/ft^3).

The last term in this relationship is a correction for supersaturated liquids, assuming an incompressible fluid.

Finally, the enthalpy for UF$_6$ vapor is given by

5-6

$$H_{UF6v} = 43.2614 + 9.21307 \times 10^{-2} T_r + 6.26265 \times 10^{-6} T_r^2 + 2951.71 T_r^{-1}$$
$$+ 3.0939 \times 10^{-3} T_r \left(Z|_{P,T} - Z|_{14.7,T} \right) \quad .$$

(5.18)

where

$Z_{P,T}$ = the compressibility factor at pressure P and temperature T.

The last term in this relationship is a compressibility correction. This term is small in the atmosphere because atmospheric pressure is always near 14.7 psia.

5.2.5 Uranium Enrichment

William's (1985) model includes correction terms for the molecular weight to account for enrichment. The correction terms are retained in the UF_6 plume model. The molecular weight of enriched uranium is input to the model along with the release rates. RASCAL 3.0.5 corrects for molecular weight, but, the correction has only a ver small effect.

5.2.6 HF-H₂O System

William's (1985) model treats HF and H_2O as a system for computation of vapor pressures and enthalpy assuming that the HF and H_2O are vapors in equilibrium with a condensed phase. It is unlikely that a condensed phase will occur in the atmosphere because of the exothermic nature of the UF_6/H_2O reaction. However, the equations for the condensed phase are included in the UF_6 plume model for completeness. HF vapor in the atmosphere is assumed to exist as a set of polymers linked by hydrogen bonding. The effects of this self association are included in the HF vapor pressure and enthalpy calculations.

5.2.7 HF Self Association

Williams (1985), Beckerdrite, Powell, and Adams (1983) report that the self association of HF is reasonably modeled by assuming equilibrium among an HF monomer $(HF)_1$, an HF trimer $(HF)_3$, and an HF hexamer $(HF)_6$. The partial pressure of HF is given by

$$P_{HF} = P_{(HF)_1} + K_3 P_{(HF)_1}^3 + K_6 P_{(HF)_1}^6 \quad ,$$

(5.19)

where the second and third terms on the right are the partial pressures of the polymers and K_3 and K_6 are equilibrium coefficients. The equilibrium coefficients have been determined experimentally by Strohmeier and Briegleb (Beckerdrite, Powell, and Adams 1983). Using these data, Williams (1985) derived the

$$K_3 = \exp\left(2.3884.0 T_r^{-1} - 51.2393\right)$$

(5.20)

following relationships to estimate the coefficients

and

$$K_6 = \exp\left(40319.6T_r^{-1} - 87.7927\right) .$$ (5.21)

With self association, the effective molecular weight for HF for vapor-phase densities and mole fractions is greater than the molecular weight of the HF monomer. It is

$$MW_{HF} = \frac{\left[P_{(HF)_1} MW_{(HF)_1} + K_3 P_{(HF)_1}^3 MW_{(HF)_3} + K_6 P_{(HF)_1}^6 MW_{(HF)_6}\right]}{P_{HF}}$$ (5.22)

5.2.8 Partial Vapor Pressures of HF

If a condensed phase exists in the HF-H_2O system, the vapor pressure of HF is calculated using relationships of the form

$$P_{HF} = \exp\left(AT_r^{-1} + B\right) ,$$ (5.23)

where

A and B = the model parameters that are a function of the weight fraction of HF in the condensed phase.

Williams (1985) gives estimates of the coefficient values based on a figure supplied by Allied Chemical (Brian C. Rogers). The differences between partial vapor pressures estimated using the model and the figure, range from about 1% for weight fractions near 1.0 to a maximum of 5% at weight fractions below 0.5. If a condensed phase does not exist, the partial vapor pressure of HF is calculated using an iterative procedure along with estimation of the effective molecular weight.

5.2.9 Partial Vapor Pressure of H_2O

Until all the UF_6 has reacted with water, all water entering the plume will be used by the UF_6/H_2O reaction to form HF and UO_2F_2. Under these conditions, the H_2O partial vapor pressure in the HF-H_2O system will be zero. Following conversion of all of the UF_6, an initial estimate is made of the H_2O partial vapor pressure from the mass of water in the plume using the ideal gas law. The phase composition of the HF-H_2O system is determined by comparing the sum of the HF partial pressure and the initial estimate of the H_2O partial pressure with the total pressure of HF and H_2O for an azeotropic mixture. If the sum is less than the total pressure for the azeotropic mixture, there is no condensation phase and the initial H_2O partial pressure estimate is used. If condensation occurs, an iterative procedure is used to determine the partial pressure of H_2O. The procedure is described in detail by Williams (1985).

5.2.10 Enthalpy of HF-H$_2$O Vapor Mixtures

The enthalpy of HF-H$_2$O vapor mixtures is given by

$$H_{HFH2Ov} = 1051.0 + 0.472T_f - \left[376.0 + 0.136T_f + 790.642W_{(HF)_3} + 667.358W_{(HF)_6} \right]W_{HFv} \, , \quad (5.24)$$

where

$W_{(HF)3}$ and $W_{(HF)6}$ = the weight fractions of the HF polymers with respect to total HF,
W_{HFv} = the weight fraction of HF in the HF-H$_2$O vapor.

The heat of association for (HF)$_3$ is -790.642 Btu/lb$_m$ of (HF)$_3$ formed, and the heat of association of (HF)$_6$ is -667.358 Btu/lb$_m$ of (HF)$_6$ formed.

$$H_{HFH2Ol} = A_i + B_iW_{HFl} + C_iW_{HFl}^2 \, , \quad (5.25)$$

5.2.11 Enthalpy of HF-H$_2$O Liquid Mixtures

The enthalpy of a liquid HF-H$_2$O mixture is given by a relationship of the form
where the coefficients A_i, B_i, and C_i are functions of the weight fraction W_{HFl} of HF in the HF-H$_2$O liquid mixture. Williams (1985) provides correlations for estimating the coefficients that are based on an enthalpy-concentration diagram provided by Brian C. Rogers at Allied Chemical.

5.2.12 UO$_2$F$_2$ Enthalpy

UO$_2$F$_2$ is formed as a product of the UF$_6$-H$_2$O reaction. It is a solid with a heat capacity of 0.0821 Btu/(lb$_m$ °F). The enthalpy at any temperature, relative to a reference temperature is

$$H_{UO2F2} = 0.0821\left(T_f - T_{ref} \right) \, , \quad (5.26)$$

where

T_f = UO$_2$F$_2$ temperature, and
T_{ref} = reference temperature (both in °F or °R).

The reference temperature is 77°F in the UF$_6$ plume model.

5.2.13 Mixture Enthalpies and Plume Temperature

Mixing and reactions in the UF$_6$ plume model are assumed to take place under constant pressure. The following reference conditions are assumed for enthalpy calculations: a pressure of 1013.25 mb (1 atmosphere, 760 mm Hg, or 14.696 psia), a temperature of 25°C (77°F); a vapor state for UF_6, H_2O, and air; monomeric vapor for HF; and solid for UO_2F_2.

The enthalpy of the plume is calculated for the control volume as the control volume moves downwind. The control volume initially consists of the volume of the UF_6 plus the volume of the entrained air and water vapor and has an enthalpy equal to the sum of enthalpies of the UF_6, air, and H_2O. With the UF_6-H_2O reaction, the enthalpy of the control volume increases because of the heat release and changes in the masses of the plume constituents.

The UF_6-H_2O reaction is limited by one constituent or the other. If the available water is the limiting factor, the heat of reaction is calculated as
where

$$H_{rxn} = 25.199 \times 10^3 \frac{m_{H2O}}{MW_{H2O}} , \qquad (5.27)$$

H_{rxn} = heat of reaction (Btu),
m_{H2O} = mass of water available for the reaction (lbm),
MW_{H2O} = molecular weight of water (lbm/lbm-mole).

Otherwise, the heat of reaction limited by the available UF_6 is calculated as

$$H_{rxn} = 50.398 \times 10^3 \frac{m_u}{MW_u} . \qquad (5.28)$$

where

m_u = mass of UF_6 available for the reaction (lbm),
MW_u = molecular weight of UF_6 (lbm/lbm-mole).

Note that the constants in Equations 27 and 27 have units of Btu/(lbm-mole).

With completion of the UF_6-H_2O reaction, the enthalpy of the plume in the control volume is

$$H_{plume} = \Delta H_{air} + \Delta H_{H2Ov} + \Delta H_{UF6} + \Delta H_{HFH2O} + \Delta H_{UO2F2} + H_{rxn} . \qquad (5.29)$$

The change in enthalpy of air is

$$\Delta H_{air} = 0.24037 m_{air} \left(T_{air} - 77.0 \right) \qquad (5.30)$$

and, the change in enthalpy associated with entrained water is

$$\Delta H_{H2O} = \left(0.99783 m_{H2Ole} + 0.472 m_{H2Ove} \right) \left(T_{air} - 77.0 \right) . \qquad (5.31)$$

where

m_{H2Ole} = mass of liquid water entrained, and

m_{H2Ove} = mass of water vapor entrained.

Finally, an iterative procedure is used to arrive at a plume temperature that gives the same mixture enthalpy. During this procedure, the phase composition of the HF-H_2O mixture and UF_6 are adjusted as the temperature changes. The convergence criterion for plume temperature is 0.1 °C. This precision is more than adequate because the plume temperature is used only in plume-rise calculations.

5.3 Dispersion and Deposition of HF and UO_2F_2

The UF_6 model works in two stages. In the first stage, the model calculates the spread of UF_6, the conversion of UF_6 to HF and UO_2F_2, and the plume rise of the HF and UO_2F_2. The products of this stage are UF_6, HF, and UO_2F_2 source terms and the plume rise of HF and UO_2F_2, all as a function of distance from the release point. In the second stage, a straight-line Gaussian model (based on the model described in Chapter 4) is used to calculate airborne concentrations and deposition of HF and UO_2F_2 at receptors on a polar grid. The distance-dependent source terms calculated in the first stage are used as long as UF_6 is present. After the UF_6 is gone, the HF and UO_2F_2 source terms are depleted to account for deposition.

The UF_6 chemical and thermodynamics models are run in the first stage while the control volume moves downwind in small time steps. The maximum time step is 15 seconds. If, with the 1-meter wind, the UF_6 control volume would reach the first arc of receptors in less than 75 seconds, the time step is reduced so that the control volume reaches the first arc at the end of fifth time step. As the control volume moves downwind, plume rise is calculated using the stable plume equations discussed in Section 4.5.2. In addition, transition plume rise is calculated using

$$\Delta h_t = 1.6 F_b^{1/3} x^{2/3} u^{-1}$$ (5.32)

where

Δh_t = transition rise (m) (Briggs 1984),
F_b = buoyancy flux (m^4/s^3),
x = downwind distance, and
u = 10-m wind speed (m/s).

The smaller of the transition and final rise is selected as the plume rise.

Dispersion of the HF and UO_2F_2 plumes is calculated using the dispersion parameters that are used in the main RASCAL 3.0.5 plume model. As long as unreacted UF_6 is present, the HF and UO_2F_2 plumes are assumed to be uniformly mixed in the vertical because the plumes are being fed by the UF_6 - H_2O reaction. Normalized HF and UO_2F_2 concentrations in this range are given by

$$\chi / Q' = \frac{1}{(2\pi)^{1/2} u \Sigma_y H} \exp\left[-\frac{1}{2}\left(\frac{y}{\Sigma_y} \right)^2 \right], \tag{5.33}$$

where

$$\Sigma_y = \left[\sigma_y^2 + \left(\frac{w_{UF6}}{4} \right)^2 \right]^{1/2} \tag{5.34}$$

and

$$H = \Delta h_r + 3\sigma_z. \tag{5.35}$$

In these last two Equations, σ_y and σ_z are the horizontal and vertical dispersion parameters for a point source plume (see Section 4.3), and w_{UF6} is the width of the UF_6 control volume.

After all UF_6 is converted to HF and UO_2F_2, their normalized concentrations are given by

$$\chi / Q' = \frac{1}{\pi u \Sigma_y \Sigma_z} \exp\left[-\frac{1}{2}\left(\frac{y}{\Sigma_y} \right)^2 \right] F(x) \tag{5.36}$$

where w_{UF6} is a constant equal to its value just before the last UF_6 is converted to HF and UO_2F_2 and

$$\Sigma_z = \left[\sigma_z^2 + \left(\frac{H}{2} \right)^2 \right]^{1/2} \tag{5.37}$$

As with w_{UF6}, H is a constant equal to its value just before the last UF_6 is converted to HF and UO_2F_2. Finally, $F_x(x)$ is the vertical distribution function described in Section 4.1.1. The receptor height, z, is assumed to be 1 meter.

5.4 Calculated Result Types

The RASCAL 3.0.5 UF_6 plume model Calculates the following result types as a function of distance.

- **Airborne uranium exposure (g-s/m³)**

  The airborne uranium exposure includes total exposure to uranium It includes contributions from both UF_6 and UO_2F_2. For this calculation only, UF_6 is assumed to be a trace gas, not a dense gas.

- **Inhaled uranium (mg)**

Inhaled uranium is calculated from the total exposure using a breathing rate passed from the user interface. The default breathing rate is 3.33×10^{-4} m³/s.

- **Committed effective dose equivalent from inhaled uranium (rem)**

The committed effective dose equivalent (CEDE) from uranium is calculated from the inhaled uranium, the specific activity of the uranium, and the inhalation dose factors from Federal Guidance Report No. 11 (Eckerman et al. 1988). In addition, all dose calculations have been eliminated except for the calculation of the inhalation CEDE for uranium. This CEDE can be considered equivalent to the TEDE since cloudshine and groundshine doses are negligible. The UF_6 plume model does not calculate cloudshine or groundshine doses because they are negligible.

- **Deposited uranium (g/m²)**

Deposited uranium includes uranium in any UF_6 that condenses before reacting with atmospheric water and uranium in UO_2F_2 that deposits from UO_2F_2 plume. UF_6 vapor does not deposit, and there is no enhanced deposition of UO_2F_2 following its formation.

- **Average HF concentration in the lung (ppm by volume)**

The HF concentration calculated by UF_6 plume model is the HF concentration in the lungs. The concentration in the lungs includes inhaled HF plus HF formed as a result of the reaction of inhaled UF_6 with water in the lungs. If no UF_6 is inhaled, the HF concentration in the lung is the same as the concentration in the atmosphere.

- **1-h equivalent HF concentration in the lung (ppm by volume)**

The 1-h equivalent HF concentration in the lung is an effective concentration calculated for short-duration releases for comparison with toxicity limits. It is calculated as

$$ C_{1he} = C(t)\left(\frac{t}{3600}\right)^{1/2} , \qquad (5.38) $$

where

C_{1he} = 1h equivalent concentration (ppm),
$C(t)$ = average concentration for duration t (ppm),
t = duration of the exposure to concentration $C(t)$ (s).

- **HF deposition (g/m²).**

HF deposition is calculated from the atmospheric HF exposure using a 0.003 m/s deposition velocity.

5.5 Comparison of RASCAL UF6 Plume Model with Experimental Measurements and Results from Other Models

The transport and dispersion portions of the UF_6 plume model have been evaluated for small UF_6 releases by comparison with measurements from three French experiments and comparison with two other UF_6 models. Figure 5.1 shows the comparisons. The experiments, data, other models are described in NUREG/CR-6481 (Nair et al. 1997). The figure compares average uranium concentrations predicted by the RASCAL 3.0.5 UF_6 plume model and the other models with average concentrations measured between 10 and 500 meters from the release point. In general, RASCAL 3.0.5 UF_6 plume model tends to over-predict the uranium concentrations by less than a factor of 2. The other models tend to over predict by larger factors.

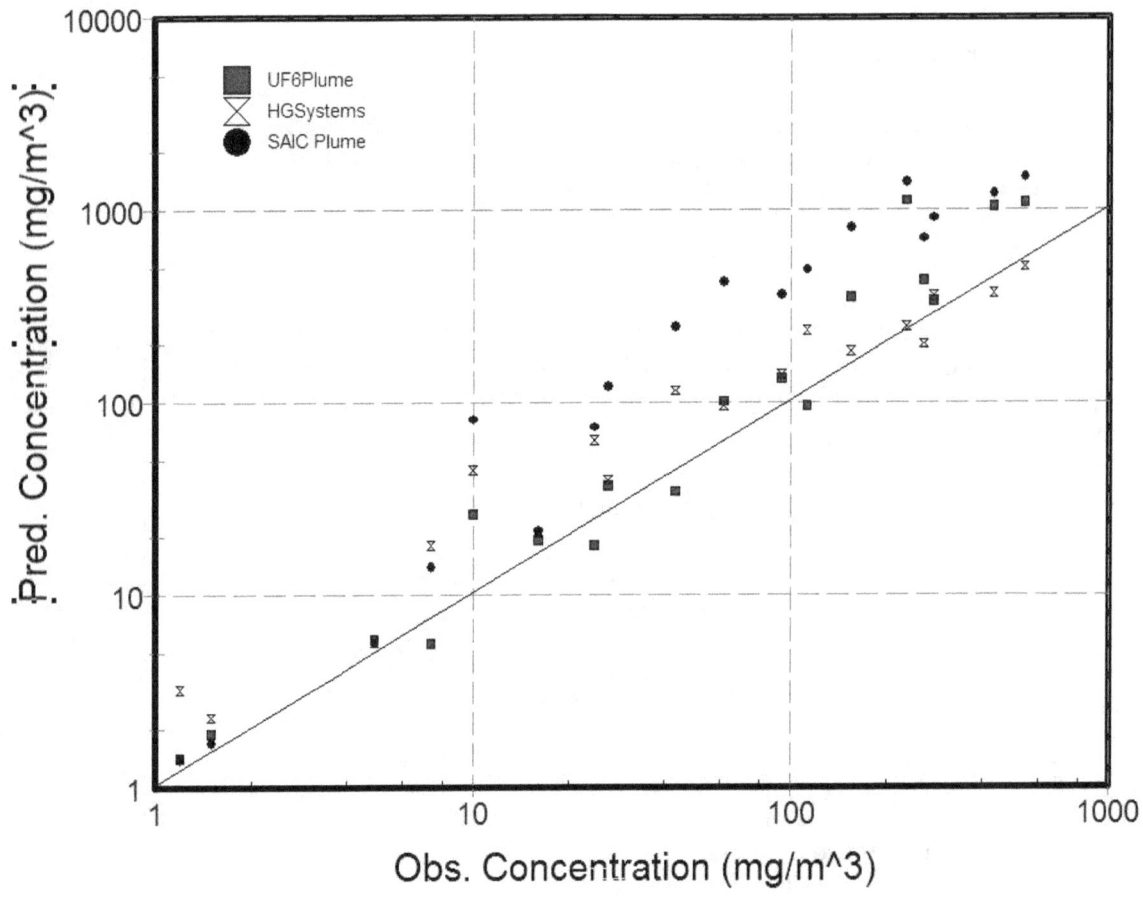

Figure 5.1 Comparison of UF6Plume model predictions of average uranium concentrations with measured concentrations and predictions of other models.

5.6 References

Beckerdrite, J. M., D. R. Powell, and E. T. Adams. 1983. "Self-Association of Gases, 2: The Association of Hydrogen Fluoride." *J. Chem. Eng. Data,* **28**:287–93.

Briggs, G. A. 1984. "Plume Rise and Buoyancy Effects." *Atmospheric Science and Power Production.* Ed. D. Randerson, DOE/TIC-27601, U. S. Department of Energy.

Dewitt, R. 1960. *Uranium Hexafluoride: A Survey of the Physico-Chemical Properties.* GAT-280, Goodyear Atomic Corporation, Portsmouth, Ohio.

Eidsvik, K.J. 1980. "A Model for Heavy Gas Dispersion in the Atmosphere." *Atmos. Environ.* **14**:769-777.

Nair, S.K., D.B. Chambers, S.H. Park, Z.R. Radonjic, P.T. Coutts, C.J. Lewis, J.S. Hammond, and F. O. Hoffman. 1997. *Review of Models Used for Determining Consequences of UF_6 Release – Model Evaluation Report.* NUREG/CR-6481 Vol, U. S. Nuclear Regulatory Commission

Williams, W. R. 1985. *Calculational Methods for Analysis of Postulated UF6 Releases.* 2 Vols., NUREG/CR-4360, U. S. Nuclear Regulatory Commission.

6 Meteorological Data Processor

The meteorological data processor is the part of RASCAL 3.0.5 that allows the user to enter meteorological data and prepares the data for use by the atmospheric transport and diffusion models. Meteorological data for the site (release point) and as many as 35 additional meteorological stations may be entered. All data must be entered manually at the present time. A future version of the meteorological data processor may include options for importing meteorological data files.

The following sections describe the technical aspects of the meteorological data requirements and the meteorological data processing.

6.1 Model Domain

Model domain refers to the area covered by the dose calculations in the source term to dose model. The model domain for the puff model is square. The release point is at the center of the square. The user can select a square that is 20, 50, or 100 miles (32, 80, and 160 km) on a side. The meteorological data processor creates a meteorological file for the selected domain for use by the puff model. This file describes the spatial and temporal variation of meteorological conditions at nodes on a Cartesian grid.

The meteorological data processor also creates a smaller file of meteorological data for the release point for use by plume model and the UF_6 plume model. This file describes the temporal variation of meteorological conditions at the center of the polar grid used by these straight-line models.

Meteorological data must be entered for at least one location, generally the release site. If data are not available for the release site, they may be entered for another location. However, the release location must still be selected as if it had meteorological data because the location of the release site is used to fix the position of the grid.

6.2 Meteorological Stations

Locations for which meteorological data are entered are called meteorological stations. Meteorological stations may be within or near the model domain. Spatial fields of winds, stability, etc., are produced from the station meteorological data. Section 6.3 discusses the meteorological data input.

Information on each station must be available before a station's meteorological data can be used in generating the meteorological fields needed by the puff model. Files containing the required information for each operating nuclear power plant, large fuel cycle facilities, and several major radioactive materials facilities are included with RASCAL 3.0.5. Each file contains the information for the location of the site and for some selected meteorological observation stations near the site.

Since the file for a specific site does all meteorological stations near the site and since the information for meteorological stations can become outdated, the meteorological data processor provides the means to add meteorological stations to the file or update information in the file.

The following information is needed for each meteorological station:

- A station identification. The station identification can be any 10 letter character; the release site ID is appropriate for the first station in the station list; FAA or ICAO location indicators are appropriate for national weather service or other stations that have them. A longer station name may be included for each station for better identification.

- The latitude (positive north of the equator and negative south of the equator) and longitude (positive east of the prime meridian and negative west of the prime meridian) of the station in decimal degrees.

- The elevation of the station in meters above mean sea level.

- The surface roughness for the station (m). Meteorological texts such as Panofsky and Dutton (1984) and Stull (1988) provide guidance on estimating surface roughness. If no other information is available, a default surface roughness of 0.2 m may be used.

- The height at which the wind measurements are made (m). The instrument height should be height above ground level.

The meteorological data processor uses the station position to place station data at the proper location in the modeling domain; station elevations are used in the potential flow model that adjusts wind fields for topographic effects (Section 6.5.1), and the surface roughness and height of wind measurement are used in calculating wind speed variation with height (Section 6.4.2).

The first station in the station file is the release point (the site). The latitude and longitude of the first station will define the coordinates of the center of the model domain grid. Distances from the center of the grid to the other stations are calculated using

$$x_{rs} = r_e \, \Delta \lambda_{rs} \cos \phi_s \qquad (6.1)$$

and

$$y_{rs} = r_e \, \Delta \phi_{rs} \, , \qquad (6.2)$$

where

x_{rs} = distance of the station east (+) or west (-) of the source (center of the grid) (km),
y_{rs} = distance of the station north (+) or south (-) of the source (center of the grid) (km),
r_e = radius of the earth (= 6370 km),
φ_s = latitude of the center of the grid (release point),
$\Delta\lambda_{rs}$ = longitude difference between the station and the source (center of the grid) (radians),
$\Delta\varphi_{rs}$ = latitude difference between the station and the source (center of the grid) (radians).

6.3 Meteorological Data Input

Station meteorological data are entered for specific dates and times. The data may be actual observations (measurements) or they may be taken from meteorological forecasts. If available, the following data should be entered for each station used:

- whether the data is an observation or a forecast
- the time of the data; (the program will round the time to the nearest quarter-hour. For example, an entry of 12:07 would be changed to 12:00. Similarly, an entry of 14:22 would be changed to 14:30)
- surface-level wind speed
- surface-level wind direction
- estimated atmospheric stability
- precipitation type
- estimated mixing height (optional).

If the plume rise option is used the ambient air temperature should also be entered.

If the calculation is for a UF_6 release, the following data should also be entered for the release point (center of the grid):

- ambient air temperature
- pressure
- humidity measurement (dew point temperature, relatively humidity, or wet bulb temperature).

The air temperature is used in plume rise calculations in all three of the transport and diffusion models. The air temperature, pressure, and humidity are used in the thermodynamic calculations in the UF_6 plume model. Pressure and humidity are used only by the UF_6 plume model.

The following sections describe the meteorological variables in more detail.

6.3.1 Surface Winds

The surface winds are made up of the wind direction (the direction from which the wind is blowing) and the wind speed. Wind directions must be entered in degrees from 0° to 360°. Wind speed can be entered in units of m/s, mph, or knots and can range from 0 to 30 m/s (or equivalent in other units).

6.3.2 Atmospheric Stability Class

Atmospheric stability may be entered either as a stability class (A - extremely unstable through G - extremely stable) based on the general classification scheme discussed by Pasquill (1961), Gifford (1961), and Turner (1964) or as temperature variation with height (dT/dz). If dT/dz is entered, it will be converted to a stability class using the conversion table shown in Table 6.1 (NRC, 1972).

Table 6.1 Estimated Pasquill-Gifford Stability Class Based on NRC Delta TM Method

Stability Class	dT/dz (°C/100 m)	dT/dz (°F/100 ft)
A	<-1.9	<-1
B	<-1.7	<-0.9
C	<-1.5	<-0.8
D	<-0.5	<-0.3
E	< 1.5	< 0.8
F	< 4	< 2.2
G	≥ 4	≥ 2.2

Reference: NRC, 1972

If the user does not enter a stability class, Table 6.2 is used to estimate the stability class from wind speed, precipitation type, and the time of day. Factors discussed by Turner (1964) were used in selecting the stability classes for the table. Daytime is defined as one hour after sunrise to one hour before sunset. Nighttime is defined as one hour before sunset to one hour after sunrise. The user may alternatively select "persistence" to determine stability class. If persistence is selected, the stability entered for the earlier time will be used.

Table 6.2 Estimated Atmospheric Stability Class for Missing Stability Classes

Wind Speed (m/s)	No or Light Precipitation	Moderate or Heavy Precipitation
Daytime		
≤ 6.0	C	C
> 6.0	D	D
Nighttime		
≤ 3.0	F	E
3.1 - 5.0	E	E
> 5.0	D	D

The meteorological data processor will also compare the entered stability class with ranges of stability classes that would be expected given the time of day and meteorological conditions and replace values that are out of the expected range with more likely values. Table 6.3 is used to determine the reasonable range of stability classes given the time of day, wind speed, and precipitation type. These ranges are based on factors discussed by Turner (1964). If a meteorological station stability class falls within the reasonable range, the stability class is not modified. But if the stability class falls outside the range, the stability class is changed to the closest stability class within the range. The user has the option of turning this option off if desired.

Table 6.3 Limits of Atmospheric Stability Classes Based on Time of Day, Wind Speed, and Precipitation

Wind Speed (m/s)	No or Light Precipitation	Moderate or Heavy Precipitation
Daytime		
≤ 3.0	A - E	C - E
3.1 - 5.0	B - D	C - D
> 5.0	C - D	C - D
Nighttime		
≤ 3.0	C - G	C - E
3.1 - 5.0	D - F	D - E
5.1 - 6.0	D - E	D - E
> 6.0	D	D

6.3.3 Precipitation Type

Wet deposition in the plume model, puff model, and UF_6 plume model and the reaction between UF_6 and water in the UF_6 plume model are affected by precipitation. Information on precipitation may be entered for each station during meteorological data entry by selecting one of seven precipitation types, or unknown, if appropriate. Precipitation types are: none; light, moderate, and heavy rain; and light, moderate, and heavy snow. Rain includes drizzle, freezing rain, and freezing drizzle. Snow includes snow grains, snow pellets, ice pellets, ice crystals, and hail. The meteorological data processor estimates precipitation rates from these precipitation types (see Section 6.4.6).

6.3.4 Mixing Height

The plume and puff models use the mixing height to limit vertical dispersion. Mixing heights may be entered with the other meteorological data for a station. However, this information is generally not available. Consequently, unless the option of entering mixing-height data is specifically selected, the meteorological data processor will estimate mixing heights from wind speed and stability. There is also an option of using climatological mixing-height estimates in place of measured or calculated values.

6.3.5 Temperature

The ambient air temperature should be entered for the release point. If the meteorological data are to be used for consequence analysis for a ground-level release not involving UF_6, the temperature may be omitted without affecting doses.

6.3.6 Pressure

The station atmospheric pressure (not sea-level pressure at the station) is needed for thermodynamic calculations in the UF$_6$ plume model. However, the calculations are not particularly sensitive to the pressure as long as the pressure is within a few percent of the actual value. The meteorological data processor includes default pressures for fuel-cycle facilities that should be adequate for most purposes because atmospheric pressures rarely vary by more than ±5%. The program will convert pressures entered in other units to millibars.

6.3.7 Humidity

Information on humidity is needed for chemical reaction and thermodynamic calculations in the UF$_6$ plume model. Humidity information may be entered for the release point as dew point temperature, relative humidity, or wet bulb temperature. The meteorological data processor includes default humidity information based on climatological data for fuel-cycle facilities. However, actual data should be entered whenever possible because the calculations are very sensitive to humidity, and humidities have a wide range of variation in the atmosphere.

6.3.8 Temporal Interpolation of Input Values

The atmospheric models in RASCAL 3.0.5 expect meteorological data on the 15-minute time interval typically used to record meteorological data at U.S. nuclear power plants. Data from other meteorological stations are not likely to be available on that interval. Consequently, the meteorological data processor will estimate missing 15-minute data for each station by linear interpolation between observed values. For example, if 10:00 and 11:00 observations are entered for a station, the program will estimate values for 10:15, 10:30, and 10:45. The program will not interpolate between observed and forecast values, or between two forecast values.

The interpolation procedures are as follows:

Winds

- If the winds for both the earlier and later observation are valid, the winds are interpolated as follows:

  1) The wind speed and direction are converted to U (east-west) and V (north-south) components.

  2) The U and V component are linearly interpolated (i.e., $U(t) = (U_l - U_e)\{(t-t_e)/(t_l-t_e)\} + U_e$ where U_l, U_e, t_l, t_e are the U component and time of the later and earlier observations, respectively).

  3) The U and V components are converted back to speed and direction.

- If the wind (either speed or direction) is missing for the later observation but not for the earlier observation, the winds are set to the wind of the earlier for times within 12 hours of the earlier observation, otherwise they are assumed to be missing.

- If the wind is missing for the earlier observation, the winds at all times between the two observations are also assumed to be missing.

Atmospheric Stability

- If both observations have valid atmospheric stabilities, the atmospheric stability is estimated using linear interpolation between the two observations. If the atmospheric stability is given by a Pasquill-Gifford stability class (1-7), then the interpolation is rounded to the nearest integer.

- If the atmospheric stability for the later observation is missing but not for the earlier observation, then the atmospheric stabilities are set to the value of the earlier observation as long as the elapsed time from the earlier observation is less than 12 hours. After 12 hours, the stabilities are set to missing.

- If the atmospheric stability for the earlier observation is missing, the atmospheric stabilities at all times between the two observations are assumed to be missing.

Precipitation Type

- If both observations have valid precipitation types, then the precipitation type for the earlier observation is used when the time is less than or equal to half way between the two observations. If the time is greater than half way between the two observations, the precipitation type of the later observation is used.

- If the precipitation type for the later observation is missing but not for the earlier observation, then precipitation types are set to the precipitation type of the earlier observation as long as the elapsed time from the earlier observation is less than 12 hours. After 12 hours, the precipitation types are set to missing.

- If the precipitation type for the earlier observation is missing, the precipitation type for all times between the two observations will be set to missing.

Mixing Height

- Unless the mixing height is being entered directly (not calculated from the meteorological data or from climatology), the method of estimating mixing heights being used for the earlier observation will continue to be used.

- If the mixing heights are being entered directly, the technique used to interpolate stability is used to estimate the missing mixing heights (see previous statement).

Temperature, Pressure, and Moisture

- The same technique previously explained is used to estimate missing stabilities is used for temperature, pressure, and humidity.

- If data are missing between observed and forecast values, persistence will be used to estimate missing values for all times up to the time of the forecast.

• For each station, data for all times before the first date and time with an entered value are assumed to be missing. Likewise, data for all dates and times past the entered value are assumed to be missing.

6.4 Other Meteorological Parameters

Meteorological data entered for a station are used to evaluate additional parameters. The following subsections describe these additional parameters.

6.4.1 Monin-Obukhov Length

The Monin-Obukhov length (L) is a scaling length for vertical motions in atmospheric boundary layer studies that is used as a measure of atmospheric stability. It is used in wind profile, turbulence, and mixing-layer depth calculations. Golder (1972) provides a graphical means for converting from Pasquill-Gifford stability classes to Monin-Obukhov lengths using the surface roughness length (Section 6.4.2). The meteorological data processor uses a procedure that was developed by Ramsdell, Simonen, and Burk (1994) based on Golder's work to convert stability classes to Monin-Obukhov lengths.

6.4.2 Wind Speed vs. Height

The RASCAL 3.0.5 atmospheric dispersion models use winds that are representative of 10 meters above ground level for ground-level release calculations and winds representative of the release height for elevated release calculations. Wind measurements are not always made at these heights. Therefore, the meteorological data processor adjusts wind speeds for the difference between the measurement height and the height required for model calculations. A diabatic wind-profile model, which accounts for the effects of surface roughness and atmospheric stability on variation of wind speed with height, is used for this adjustment. No attempt is made to model the variation of wind direction with height.

The diabatic profile model is derived from atmospheric boundary layer similarity theory proposed by Monin and Obukhov (1954). The basic hypothesis of similarity theory is that a number of parameters in the atmospheric layer near the ground, including wind profiles, should be universal functions of the friction velocity, a length scale, and the height above the ground. The length scale is referred to as the Monin-Obukhov length and the ratio z/L is related to atmospheric stability.

The diabatic wind profile is

$$u(z) = \frac{u_*}{k}\left[\ln\left(\frac{z}{z_0}\right) - \psi\left(\frac{z}{L}\right)\right] , \qquad (6.3)$$

where

$u(z)$ = wind speed at height z (m/s),
u_* = friction velocity (boundary-layer scaling velocity) (m/s),
k = von Karman constant (≈ 0.4),
z_0 = surface roughness length (m),
$\psi(z/L)$ = stability correction factor,
L = Monin-Obukhov length (m).

The surface roughness length is associated with small-scale topographic features. It arises as a constant of integration in the derivation of the wind profile equations and is used in several boundary-layer relationships. Texts on atmospheric diffusion, air pollution and boundary-layer meteorology (Panofsky and Dutton 1984, Stull 1988) contain tables that give approximate relationships between surface roughness and land use, vegetation type, and topographic roughness.

The term $\Psi(z/L)$ accounts for the effects of stability on the wind profile. In stable atmospheric conditions, $\Psi(z/L)$ has the form $-\alpha z/L$ where α has a value of 5. In neutral conditions $\Psi(z/L)$ is equal to zero, and the diabatic profile simplifies to a logarithmic profile.

In unstable air, $\Psi(z/L)$ is more complicated. According to Panofsky and Dutton (1984), the most common form of $\Psi(z/L)$ for unstable conditions, based on the work of Businger et al. (Paulson 1970) is

$$\psi\left(\frac{z}{L}\right) = \ln\left(\left[\frac{(1+x^2)}{2}\right]\left[\frac{(1+x)}{2}\right]^2\right) - 2\tan^{-1}(x) + \frac{\pi}{2}$$ (6.4)

where

$x = (1-16z/L)^{1/4}$.

Equation (6.4) is used to estimate the friction velocity (u_*) from the wind speed, surface roughness, and Monin-Obukhov length. In unstable and neutral conditions, the use of Eq. (6.4) is limited to the lowest 100 meters of the atmosphere. In stable conditions, the upper limit for application of Eq. (6.4) is the smaller of 100 meters or three times the Monin-Obukhov length.

6.4.3 Mixing Height

Heating of the surface and surface friction combine to generate turbulence that mixes material released at or near ground level through a layer that varies in thickness from a few meters to a few kilometers. This layer is referred to as the mixing layer. The atmospheric models in RASCAL 3.0.5 use the mixing height (also referred to as the mixing-layer depth and mixing-layer thickness) to limit vertical diffusion.

The meteorological data processor has three methods for obtaining estimates of the mixing height at meteorological stations. The mixing height may be entered directly, or it may be estimated by the program from either current meteorological data or climatological information. Of the latter two options, estimation of mixing height from current meteorological data is preferable to estimating the mixing height from climatological data, if sufficient data are available.

The meteorological data processor uses algorithms developed by Ramsdell, Simonen, and Burk (1994) for estimating mixing height from current meteorological data. The algorithms are based relationships derived by Zilitinkevich (1972) for stable and neutral conditions.

For stable atmospheric conditions, the relationship is

$$H = k\left(\frac{u_* L}{f}\right)^{1/2},$$ (6.5)

where

H = mixing height (m),
k = von Karman constant (0.4),
u_* = friction velocity (m/s),
L = Monin-Obukhov length (m),
f = Coriolis parameter (1/s).

A 50-meter mixing height is used if the mixing height calculated by Equation 6.5 is less than 50 meters. Similarly, if the calculated mixing height is greater than 2,000 meters, the mixing height is set to 2,000 meters.

For neutral and unstable conditions, the mixing height is calculated by

$$H = \frac{\beta u_*}{f} \; , \tag{6.6}$$

where

β is a constant set to 0.2.

If the mixing height calculated by Equation 6.6 is less than 250 meters, the mixing height is set to 250 meters, and if the calculated mixing height is greater than 2,000 meters, the mixing height is set to 2,000 meters.

The mixing-layer thickness may also be estimated from climatological data. When this option is selected, the mixing-layer thickness is estimated from typical morning and afternoon thicknesses for each month using the method used in the Environmental Protection Agency's meteorological preprocessor code, PCRAMMET (EPA 1999). The monthly morning and afternoon mixing-layer thicknesses were calculated from daily data obtained from the Environmental Protection Agency's Support Center for Regulatory Air Models (www.epa.gov/scram001/). The following rules are used in estimating mixing-layer thicknesses from the monthly values.

• From midnight to sunrise - use the morning mixing height.

• From sunrise to 1400 - linearly interpolate between morning and afternoon mixing heights.

• From 1400 to sunset - use the afternoon mixing height.

• From sunset to midnight - use exponential interpolation between the afternoon and morning mixing heights. For the last day of the month, use the morning of the next month.

The exponential interpolation of the mixing height is given by

$$H(t) = a \cdot \exp\left(-\frac{bt}{24}\right) \; , \tag{6.7}$$

where

$H(t)$ = mixing height at time t (m),
$a = H_{morn}/\exp(-b)$,
$b = 24 \ln(H_{aft}/H_{morn})(24 - t_{sunset})$,
H_{morn} = morning mixing height (m),
H_{aft} = afternoon mixing height (m),
t_{sunset} = time of sunset (h).

The following equations (Stull 1988) are used in calculating sunset and sunrise times

$$\sin \upsilon = \sin \phi \sin \delta_s - \cos \phi \cos \delta_s \cos(T_0) \ , \tag{6.8}$$

where

υ = local elevation of the sun,
φ = latitude of the station,
δ_s = solar declination angle (angle of the sun above the equator),
T_0 = local time.

The solar declination angle is calculated using

$$\delta_s = \phi \cos\left(\frac{2\pi d - d}{d}\right) \ , \tag{6.9}$$

where

φ_r = latitude of the tropic of Cancer (23.45°),
d = Julian calendar day of the year,
d_r = Julian calendar day of the summer solstice (173),
d_y = average number of days per year (365.25).

The local time is defined as

$$T_0 = \left(\frac{\pi t_{utc}}{12} - \lambda_e\right) \ , \tag{6.10}$$

where

t_{utc} = time at the prime meridian,
λ_e = longitude (in radians) of the station.

Sunrise and sunset are calculated by setting the solar elevation angle to -0.833° (the sun appears to rise and set when it is 0.833° below the horizon) and solving for T_0 using Eq. (6.10). Sunset is 24 h T_0. The equations for sunrise and sunset do not take into account the ellipticity of the earth's orbit, but it is accurate to about ±16 minutes.

6.4.4 Dry Air and Water Vapor Density

The UF_6 plume model requires estimates of the air density and water content of the atmosphere. These two variables are estimated from the air temperature, station pressure, and humidity for the release location. (The release location is assumed to be at the location of the first meteorological station. The method used to calculate the dry air density and water-vapor density depends upon the variables used for humidity and whether the temperature, pressure, and moisture variables exist.

The water vapor density is given by

$$\rho_v = \frac{e}{R_v T} ,$$

(6.11)

where

ρ_v = water vapor density (kg/m^3),
e = vapor pressure (Pa),
R_v = gas constant = 461.5 J/kg °K,
T = temperature (°K),

and the dry air density can be estimated by

$$\rho_d = \frac{(p - e)}{RT} ,$$

(6.12)

where

ρ_d = dry air density (kg/m^3),
p = total station pressure (Pa),
R = gas constant for dry air = 287.0 J/kg °K.

The method used to calculate the vapor pressure e depends upon the moisture variable. If the moisture variable is the dew point and precipitation is not occurring, then the vapor pressure is given by

$$e = e_s(T_d) ,$$

(6.13)

where

$e_s(T_d)$ [mb] is the saturation vapor pressure at temperature T_d (°C).

According to Rogers and Yau (1989), the saturation vapor pressure is given by

$$e_s(T) = 6.112 \exp\left(17.67 \frac{T}{T + 243.5} \right) .$$

(6.14)

If the moisture is defined by the relative humidity and precipitation is not occurring, then the vapor pressure is given by

$$e = \frac{RH \cdot e_s(T)}{100} \, ,$$ (6.15)

where

RH = relative humidity (percent),

$e_s(T)$ = Eq. (6.14).

If the moisture is given by the wet bulb temperature and precipitation is not occurring, then the vapor pressure is given by

$$e = e_s(T) - \left(\frac{p}{0.622}\right) \cdot \left[\frac{1004(T - T_w)}{2.5E6}\right] \, ,$$ (6.16)

where

T_w is the wet bulb temperature [°C].

If precipitation is occurring, the air is assumed to be 95% saturated, so the vapor pressure is given by

$$e = 0.95 e_s(T) \, ,$$ (6.17)

where

$e_s(T)$ is given by Eq. (6.14).

During precipitation, Eq. (6.17) is used regardless of the moisture variable.

If the temperature, pressure, or moisture variable is missing, the climatological values are used for the dry air and water vapor density. If precipitation is occurring, the two densities will be based on Eqs. (6.12), (6.14), and (6.17) with the temperature obtained from the climate file. If the climatological values are missing, then the dry air density is assumed to be 1.2 kg/m^3, and the water vapor density is assumed to be zero.

6.4.5 Precipitation Rate

The RASCAL 3.0.5 atmospheric codes use precipitation rate to calculate the wet deposition rate for UF$_6$ releases. When the precipitation type for a station is other than none or unknown, the meteorological data processor estimates a precipitation rate (mm/h) for the station using the precipitation type and a precipitation rate zone. Each site in the RASCAL database is assigned to one of three precipitation rate zones. The precipitation zones, originally defined in Ramsdell, Simonen, and Burk (1994) are based on annual precipitation. Zone 1 is for areas where the annual precipitation is less than 10 inches, zone 2 is for areas where the annual precipitation is between 10 and 20 inches, and zone 3 is for areas where the annual precipitation exceeds 20 inches. Most existing reactor sites are assigned to precipitation rate zone 3. Some sites in the drier regions of the United States are assigned to zone 1. Precipitation rate zone assignments are made in the climatology database and can be modified as appropriate.

The precipitation rates assigned by the meteorological data processor are listed in Table 6.4. These rates, based on data collected in the Pacific northwest, should be conservative for most nuclear facilities in the United States.

Table 6.4 Precipitation Rates as a Function of Precipitation Climate Zone

Precipitation Type	Precipitation Rate (mm/h)		
	Zone 1	Zone 2	Zone 3
Light rain	0.4	0.6	0.7
Medium rain	3.8	3.8	3.8
Heavy rain	3.8	3.8	8.5
Light snow	0.3	0.3	0.7
Medium snow	1.7	1.7	3.8
Heavy rain	1.7	1.7	3.8

6.5 Calculating Spatially Varying Meteorological Conditions

The puff model takes into account both spatial and temporal variations in the atmospheric conditions. The meteorological data processor provides the gridded fields of the atmospheric stability class, the inverse Monin-Obukhov length, the east-west (U) and north-south (V) components of the wind, the mixing height, and the precipitation type and precipitation rate for each of the three puff model domains. The following subsections describe the preparation of the fields from the station data.

6.5.1 Wind Fields

The puff model uses wind fields to calculate movement of puffs. These are fields of U (east-west) and V (north-south) components of the wind vector. The wind fields are created from station wind data using a $1/r^2$ interpolation scheme, where r is the distance from the grid point to the station. This interpolation scheme, which was used in earlier NRC codes such as MESOI (Ramsdell, Athey, and Glantz 1983) and MESORAD (Scherpelz, et al. 1986; Ramsdell et al. 1988), is common in spatial interpolation of the wind fields (Hanna, Briggs, and Hosker 1982).

6.5.2 Adjustment of Wind Fields for Topography

If the meteorological stations reporting data are well placed with respect to major topographic features, the wind fields developed by interpolation will give reasonable puff trajectories. However, with one meteorological station or a small number of stations, the wind fields may not properly reflect the effects of topography. The meteorological data processor includes an option to use a simple one-layer model to adjust wind fields for topography. Wind field adjustments are greatest for stable atmospheric conditions (E, F, and G stability classes) and least for neutral conditions (stability class D). Wind fields are not adjusted in unstable atmospheric conditions (stability classes A, B, and C). For this purpose only, atmospheric stability at the release point (center of the model domain) is assumed to apply to the entire domain.

The wind-field model in RASCAL 3.0.5 is a two-dimensional adaption of the wind fitting program described by Ross et al. (1988) that is used in the NUATMOS and MATTHEW codes. In the RASCAL 3.0.5 implementation, wind fields created by interpolation are used as the starting point in the adjustment process. The thickness of the mixing layer is calculated for each node in the model domain by computing the difference between the top of the boundary layer and the terrain elevation. For those nodes where the terrain rises above the top of the boundary layer, the program assumes that the boundary-layer thickness is 0.01 meter and sets the wind to zero. This technique is simple to implement and has proven effective at generating flows that avoid obstacles such as mountain ridges.

The initial wind field is then adjusted using methods of variational calculus to produce a non-divergent wind field in the boundary layer, subject to the constraint of minimum difference between the initial wind field and the adjusted wind field. The procedure for adjusting the wind field involves solving Poisson's equation. The code uses a nine-point Laplacian operator and a simultaneous relaxation technique to obtain the solution.

The model has been tested and shows that the winds produced by the model flow around obstacles that are well resolved by the grid. Obstacles having width of three grid points or greater are considered well resolved. Smaller obstacles may or may not be resolved, depending on their shape and orientation relative to the grid. For example, a ridge one grid point wide is well resolved if it runs in the x or y direction, but if that same ridge is at 45° to the grid, it is not resolved.

The adjusted wind field is most accurate near stations and along trajectories that pass near stations. Wind fields are less accurate elsewhere. Thus, it is desirable to have wind data near the release point and, if possible, at downwind locations.

Topographic data are included in RASCAL 3.0.5 for all sites in the database. Sites not included in the RASCAL database and the generic site do not have topographic data files. Therefore, the option of modifying wind fields for topographic effects is not available for these sites.

6.5.3 Stability and Precipitation

The stability class and precipitation fields (precipitation type and precipitation rate) are based on data for the closest meteorological station. Fields created in this manner include stability class, inverse Monin-Obukhov length, precipitation type, and precipitation rate. This procedure avoids averaging that would minimize the effects of extreme stability or instability. It also provides maximum detail in treating isolated precipitation events.

6.5.4 Mixing Height

Estimates of station mixing height are not considered particularly reliable. Therefore, the spatial variation of the mixing height is modeled using two steps. The initial step is to create a mixing height field using the mixing height for the closest station for each point in the field. If there is only one station the process is terminated after this step. The second step is taken when there are two or more stations. In this step, the mixing height at each point in the field is replaced by an average of the initial mixing height and the mixing heights at 24 surrounding points. This second step smooths the mixing height field.

6.6 Calculating Meteorological Conditions at the Source

All of the RASCAL 3.0.5 atmospheric dispersion models require information about the wind speed, wind direction, atmospheric stability, precipitation type, precipitation rate, mixing-layer depth, and temperature at the source. If these meteorological data are available for the release point, which is considered to be at the source, then those data are used. If no data are available for the release point, the wind speed, wind direction, atmospheric stability, current weather, precipitation rate, and mixing-layer depth will be estimated from the spatial meteorological data field. For the temperature, a default climatological value will be used if it exists. The default climatological temperature varies by month and is obtained from the climate file. If the climate file does not exists for the site, the temperature will be flagged as a missing value.

6.7 References

Gifford, F. A. 1961. "Use of Routine Meteorological Observations for Estimating Atmospheric Dispersion." *Nuclear Safety*, **2**(4):47–51.

Golder, D. 1972. "Relations Among Stability Parameters in the Surface Layer." *Boundary-Layer Meteorology*, **3**(1):47–58, 1972.

Hanna, S. R., G. A. Briggs, and R. P. Hosker. 1982. *Handbook on Atmospheric Diffusion*. DOE/TIC-11223, U.S. Department of Energy.

Monin, A. S., and A. M. Obukhov. 1954. "Basic Laws of Turbulent Mixing in the Ground Layer of the Atmosphere." *Trans. Geophys. Inst. Akad. Nauk*, USSR, **151**:163–87.

Panofsky, H. A., and J. A. Dutton. 1984. *Atmospheric Turbulence*. J. Wiley & Sons, New York.

Pasquill, F. 1961. "The Estimation of the Dispersion of Windborne Material." *The Meteorological Magazine*, **90**:33–49.

Paulson, C. A. 1970. "The Mathematical Representation of Wind Speed and Temperature Profiles in the Unstable Atmospheric Surface Layer." *J. of Applied Meteorology*, **9**:1884–89.

Ramsdell, Jr., J. V., G. F. Athey, and C. S. Glantz. 1983. *MESOI Version 2.0: An Interactive Mesoscale Lagrangian Puff Dispersion Model With Deposition and Decay*. NUREG/CR-3344, U.S. Nuclear Regulatory Commission.

Ramsdell, Jr., J. V., et al. 1988. *The MESORAD Dose Assessment Model, Volume 2: Computer Code.* Vol. 2., NUREG/CR-4000, U.S. Nuclear Regulatory Commission.

Ramsdell, Jr., J. V., C. A. Simonen, and K. W. Burk. 1994. *Regional Atmospheric Transport Code for Hanford Emission Tracking (RATCHET)*. PNWD-2224 HEDR, Battelle, Pacific Northwest Laboratories, Richland, Wash.

Rogers, R. R., and M. K. Yau. 1989. *A Short Course in Cloud Physics.* Pergamon Press, New York.

Ross, D. G. 1988. "Diagnostic Wind Field Modeling for Complex Terrain: Model Development and Testing." *J. of Applied Meteorology,* **27**:785–96.

Scherpelz, R. I., et al. 1986. *The MESORAD Dose Assessment Model.* Vol. 1., NUREG/CR-4000, U.S. Nuclear Regulatory Commission.

Snedecor, G. W., and W. G. Cochran. 1980. *Statistical Methods.* 7th Edition, Iowa State University Press, Ames, Iowa.

Stull, R. B. 1988. *An Introduction to Boundary Layer Meteorology.* Kluwer Academic Publishers, Dordrecht, Netherlands.

Turner, D. B. 1964. "A Diffusion Model for an Urban Area." *J. of Applied Meteorology,* **3**(1):83–91.

U.S. Environmental Protection Agency (EPA), 1995. *PCRAMMET User's Guide,* Office of Air Quality Planning and Standards, Emissions, Monitoring, and Analysis Division, Research Triangle Park, N.C.

U.S. Environmental Protection Agency (EPA). 1999. *PCRAMMET User's Guide.* EPA-454/B-96-001 (Revised June 1999), Research Triangle Park, N.C.

U.S. Nuclear Regulator Commission (NRC). 1972. "Onsite Meteorological Programs." Regulatory Guide 1.23.

Zilitinkevich, S. S. 1972. "On the Determination of the Height of the Ekman Boundary Layer." *Boundary-Layer Meteorology*, **3**(2):141–5.

7 Intermediate Phase Dose Calculations

This chapter describes the intermediate phase dose calculations in RASCAL 3.0.5. The intermediate phase of a radiological emergency begins after the release has terminated. Dose calculations for the intermediate phase are done to determine if the concentrations of radionuclides on the ground are likely to cause doses to residents that would be in excess of the intermediate phase protective action guides established by the U. S. Environmental Protection Agency (EPA 1992). Those protective action guides are 2 rem for the first year, 0.5 rem for the second year, and 5 rem for the entire 50 years following the event.

In the Field Measurements to Dose module of RASCAL 3.0.5, the user enters the ground concentrations of deposited radionuclides at a location. Then, RASCAL 3.0.5 calculates the intermediate phase doses for the first year, the second year, and the cumulative dose over 50 years. The doses can be calculated with or without a delay in reentry into the contaminated area. The reentry delay can be any number of days between 0 to 730 days (2 years).

In addition, RASCAL 3.0.5 calculates "derived response levels" or "DRLs." A derived response level is a measurable quantity that indicates that the deposited activity could result in an intermediate phase dose equal to one of the intermediate phase protective action guides. One type of DRL is the closed window (gamma) dose rate in mR/hour equal to the first, second, or 50 year intermediate phase PAG. The other DRL is the ground concentration of a marker radionuclide equivalent to a PAG.

The DRLs are calculated for reentry delay times ranging from 0 days to 100 days. When significant quantities of short-lived radionuclides are present, the DRLs will be different depending on how long after the release the measurement is made. In addition, the assumed reentry delay will change the DRLs.

7.1 Intermediate Phase Doses

The intermediate phase doses calculated by RASCAL are the sums of three components: external dose from groundshine, internal dose from the inhalation of resuspended particles, and internal dose from the inadvertent ingestion of surface contamination.

Inadvertent ingestion is not included as a pathway in the EPA Protective Action Guide manual nor in the assessments done by the Federal Radiological Monitoring and Assessment Center (FRMAC). Therefore, the default intake for inadvertent ingestion in RASCAL is zero. However, the RASCAL user can include inadvertent ingestion by selecting an intake larger than zero.

7.1.1 Groundshine Doses

RASCAL calculates groundshine doses $D_{gs}(T)$ from the surface concentration for occupancy time intervals T. The calculation is

$$D_{gs}(T) = GRF \times \sum_{i=1}^{N} \left[C_{gi} \times DCP_{gsi}(T) \right] \qquad (7.1)$$

where

$D_{gs}(T)$ = groundshine dose (rem) for occupancy time interval T,

GRF = ground roughness factor with a default value of 0.82 (dimensionless),

C_{gi} = initial (t = 0) ground surface concentration (Ci/m^2) of radionuclide i,

$DCP_{gsi}(T)$ = intermediate phase groundshine dose conversion parameter [rem/(Ci/m^2)] for radionuclide i and occupancy time interval T.

The occupancy time intervals T over which groundshine dose $D_{gs}(T)$ is calculated correspond to the time intervals for the intermediate phase in the EPA Protective Action Guide manual, specifically, for the case of no reentry delay: zero to one year, one year to two years, and zero to 50 years. If the RASCAL user specifies a reentry delay, the occupancy time interval then starts at the specified time of reentry.

In previous versions of RASCAL, the ground roughness factor GRF was 0.7. However, more recent measurements (Anspaugh et al. 2002), based in large part on Chernobyl data, suggest that 0.82 is a more realistic value. FRMAC has also recently adopted a GRF of 0.82.

The intermediate phase groundshine dose conversion parameters $DCP_{gsi}(T)$ are calculated as:

$$DCP_{gsi}(T) = \sum_{n=0}^{N} \left[E_{egn}(T) \times EDC_n \right]$$ (7.2)

where

$E_{egn}(T)$ = effective time of exposure to the parent nuclide ($n = 0$) or daughter ($n \geq 1$) for occupancy time interval T accounting for decay, ingrowth, and weathering [s]

EDC_n = effective dose rate coefficient for exposure to contaminated ground surface based on Table III.3 of FGR 12 [Rem / (Ci s m^{-2})]

The summation from $n = 0$ to $n = N$ is over the parent and all daughters.

The effective dose rate coefficient for exposure to contaminated ground surface EDC_n is equal to the effective dose rate coefficient in Federal Guidance Report (FGR) No. 12 (Eckerman and Ryman 1993) except when a nuclide has a short-lived daughter. For about 40 radionuclides with short-lived daughters, the parent-daughter combination is treated as a single nuclide with the parent's half-life. The effective dose rate coefficients EDC_n assigned to the combination of the parent and implicit daughter is the sum of the parent's effective dose rate coefficient from FGR 12 plus the daughter's effective dose rate coefficient (corrected for branching ratio). The treatment of these combinations of parent plus implicit daughter is discussed in more detail in Section 7.2.

The intermediate phase effective exposure duration $E_{egn}(T)$ is for a specific exposure interval T and accounts for the radiological decay of the parent radionuclide and the ingrowth and decay of daughter radionuclides on the ground during the interval. It also accounts for weathering during the exposure interval.

The groundshine effective exposure duration $E_{egn}(T)$ for each radionuclide n is

$$E_{egn}(T) = A_0^{-1} \int_0^T A_n(t) W_s(t) dt \qquad (7.3)$$

where

    A_0 = initial activity of the parent (Ci)
    $A_n(t)$ = activity of the radionuclide n (Ci)
    $W_s(t)$ = groundshine weathering function (dimensionless)

The activity of the parent radionuclide as a function of time is given by the usual exponential relationship

$$A(t) = A_0 \exp(-\lambda t) \qquad (7.4)$$

where

    A_0 = initial activity, and
    λ = decay constant (s^{-1}).

Activities for daughters are given by the Bateman equations (Benedict et al. 1987) modified to be appropriate for activity rather than atoms (Strenge 1997) and to include branching fractions. These equations are described in Section 7.2.

The groundshine weathering function $W_s(t)$ (Anspaugh et al. 2002), based in large part on Chernobyl data, is a sum of two exponential terms

$$W_s(t) = C_1 e^{-\alpha t} + C_2 e^{-\beta t} \qquad (7.5)$$

where

    t = the time after deposition in days.

The first term describes the weathering during the first few years after deposition and the second term describes long term weathering. The values of C_1 and C_2 are 0.4 and 0.6, respectively, and the values of α and β are 1.26×10^{-3} d^{-1}, and 3.8×10^{-5} d^{-1}, respectively. FRMAC has also recently adopted this weathering function.

When Equations (7.4) and (7.5) are substituted into Equation (7.3), the resulting integral can be solved in closed form. Equation (7.3) can also be integrated for daughters. These solutions, which are not shown here, are used in intermediate phase groundshine dose calculations.

7.1.2 Inadvertent Ingestion Doses

Calculation of inadvertent ingestion doses is similar to the calculation of groundshine doses. It is as follows

$$D_{ing}(T) = ING \times \sum_{i=1}^{n} \left[C_{gi} \times DCP_{ingi}(T) \right] \qquad (7.6)$$

where

$D_{ing}(T)$ = inadvertent ingestion dose (rem) for occupancy time interval T,
ING = inadvertent ingestion rate (m²/day),
C_{gi} = ground surface concentration (Ci/m²) of radionuclide i
$DCP_{ingi}(T)$ = intermediate phase inadvertent ingestion dose conversion
parameter [rem /(Ci day⁻¹)]

The intermediate phase inadvertent ingestion dose conversion parameter $DCP_{ingi}(T)$ is calculated essentially the same as for groundshine in Equation 7.2.

$$DCP_{ingi}(T) = \sum_{n=0}^{N} \left[E_{eingn}(T) \times EDC_n \right] \qquad (7.7)$$

The effective dose rate coefficients EDC_n for inadvertent ingestion are based on Table 2.2 of Federal Guidance Report No. 11 (Eckerman et al 1988), and may include implicit daughters as was discussed in Section 7.1.1 for groundshine doses.

The effective exposure duration $E_{e\ ingn}(T)$ includes decay, ingrowth and weathering in the same manner as described for groundshine above. The inadvertent ingestion weathering function used in RASCAL is the same as that used for groundshine. The same weathering factor is used because no data could be found for weathering applied to the inadvertent ingestion pathway. This approach is probably conservative because removable radioactive materials on the most commonly touched surfaces are likely to be depleted quickly.

For reasonable inadvertent ingestion rates ING (zero to 10^{-3} m²/d), intermediate phase inadvertent ingestion doses are generally small compared to groundshine doses.

7.1.3 Inhalation Doses

Intermediate phase inhalation doses are generally smaller than either intermediate phase groundshine doses or inadvertent ingestion doses. The doses are calculated assuming that activity on the ground is resuspended and then inhaled. The calculation is

$$D_{inh}(T) = V_b \times R_{so} \times \sum_{i=1}^{n} \left[C_{gi} \times DCP_{inh}(T) \right] \qquad (7.8)$$

where

$D_{inh}(T)$ = inhalation dose (rem) for occupancy time interval T,

V_b = breathing rate (m³/s),

R_{so} = initial resuspension factor (m⁻¹)

C_{gi} = ground surface concentration (Ci/m²) of radionuclide i

$DCP_{inhi}(T)$ = inhalation intermediate phase dose conversion parameter [rem-s/Ci].

The default breathing rate V_b is 2.67×10^{-4} m³/s, which represents a long term breathing rate that includes both waking and sleeping breathing rates.

The inhalation intermediate phase dose conversion parameter $DCF_{inhi}(T)$ includes decay and ingrowth and the time-dependent (weathering) portion of the resuspension function.

For inhalation, the intermediate phase dose conversion parameters include decay, ingrowth, and resuspension. These dose conversion parameters are calculated as
where

$$DCP_{inhi}(T) = \sum_{n=0}^{N}\left[E_{ean}(T) \times ECF_n\right] \tag{7.9}$$

$DCP_{inhi}(T)$ = intermediate phase dose conversion parameter for the parent radionuclide

$E_{ean}(T)$ = effective exposure duration to airborne activity of the parent nuclide (n = 0) or daughters (n≥1) during T accounting for decay, ingrowth, and resuspension

ECF_n = effective dose factor based on Table 2.1 in FGR 11 including implicit daughters as previously described.

The summation is over the parent and all daughters.

The effective exposure duration to airborne activity, $E_{ean}(T)$ for each radionuclide (parent and daughters) is

$$E_{ean}(T) = A_o^{-1}\int_0^T A_n(t)R_s(t)dt \tag{7.10}$$

where

$A_n(t)$ = activity of the radionuclide

$R_s(t)$ = time-dependent resuspension factor

RASCAL implements the resuspension factor model in NCRP Report No. 129 (NCRP 1999). For the first day, the initial resuspension factor is constant with a default value of 10^{-6} m⁻¹. Beyond 1,000 days, the resuspension factor is also constant with a value of 10^{-9} m⁻¹. For days 1 through 1,000, the resuspension factor $R_s(t)$ is modeled as

$$R_s(t) = R_{so} / t \qquad (7.11)$$

where

R_{so} = initial resuspension factor with a default value of 10^{-6}/m

t = time in days

When expressions for the activity and the resuspension factor are substituted into Equation 7.11, the effective exposure cannot be calculated by closed form integration. In RASCAL, the effective exposure duration for intermediate phase inhalation doses is estimated using numerical integration.

7.1.4 Precalculated Intermediate Phase Dose Conversion Parameters

The RASCAL database contains precalculated dose conversion parameters for groundshine, inadvertent ingestion, and inhalation for the first 30 days, the first year, the second year, and 50 years for 219 radionuclides. The precalculated dose conversion parameters reduce computing time and facilitate comparison with hand calculations.

7.1.5 Dose Conversion Parameter Calculations with Delayed Reentry

Dose conversion parameters for delayed reentry are calculated as the difference between the factor without delayed reentry and the factor for the delay period. For example, the first year dose conversion parameter for a 30-day delay in reentry is calculated as

$$DCP(30d \text{ to } 1y) = DCP(0 \text{ to } 1y) - DCP(0 \text{ to } 30d) \qquad (7.12)$$

where

$DCP(30\,d \text{ to } 1\,y)$ = dose conversion parameter over the interval from 30 days to 1 year

$DCP(0 \text{ to } 1\,y)$ = dose conversion parameter for the first year

$DCP(0 \text{ to } 30\,d)$ = dose conversion parameter for the first 30 days

Intermediate phase dose conversion parameters for the first year $D(0 \text{ to } 1\,y)$, the second year $D(1 \text{ to } 2\,y)$, and 50 years $D(0 \text{ to } 50\,y)$ have been precalculated. Intermediate phase dose conversion parameters for the delay period are calculated as needed, after the delay period is defined.

7.2 Decay and Ingrowth

In its calculation of intermediate phase doses, RASCAL accounts for decay and ingrowth using simplified decay chains. In these chains, short-lived daughters are treated implicitly assuming that they are in equilibrium with long-lived parents. Table 7.1 lists the decay chains that involve implicit daughters along with the implicit daughter and their branching fraction. Effective dose rate coefficients for the implicit daughters are added to the effective dose rate coefficient for the parent after accounting for branching fractions.

Table 7.1 Radionuclides with Implicit Daughters Assumed in Intermediate Phase Dose Calculations.

Parent	Implicit Daughters (Branching Fractions)	Parent	Implicit Daughters (Branching Fractions)
Zn-69m	Zn-69 (0.9997)	I-135	Xe-135m (0.154)
Ge-68	Ga-68 (1.0)	Cs-137	Ba-137m (0.947)
Br-83	Kr-83m (1.0)	Ce-144	Pr-144m (0.0178), Pr-144 (1.0)
Sr-91	Y-91m (0.578)	Pb-211	Bi-211 (1.0), Po-211 (0.0027), Tl-207 (0.9973)
Zr-93	Nb-93m (1.0)	Pb-212	Bi-212 (1.0), Po-212 (0.6407), Tl-208 (0.3593)
Zr-97	Nb-97m (0.947) , Nb-97 (0.053)	Bi-212	Po-212 (0.6407), Tl-208 (0.3593)
Mo-99	Tc-99m (0.876)	Rn-220	Po-216 (1.0)
Ru-103	Rh-103m (0.99974)	Rn-222	Po-218 (1.0), Pb-214 (1.0), Bi-214 (1.0), Po-214 (0.9998)
Ru-106	Rh-106 (1.0)	Ra-223	Rn-219 (1.0), Po-215 (1.0), Pb-211 (1.0), Bi-211 (1.0), Po-211 (0.0028), Tl-207 (0.9972)
Pd-103	Rh-103m (0.99974)	Ra-224	Rn-220 (1.0), Po-216 (1.0)
Ag-110m	Ag-110 (0.0133)	Ra-226	Rn-222 (1.0)
Cd-115	In-115m (1.0)	Ra-228	Ac-228 (1.0)
In-114m	In-114 (0.957)	Ac-225	Fr-221 (1.0), At-217 (1.0), Bi-213 (1.0), Po-213 (0.9784), Tl-209 (0.0216), Pb-209 (1.0)
Sn-113	In-113m (1.0)	Ac-227	Fr-223 (0.0138)
Sn-126	Sb-126m (1.0)	Th-234	Pa-234m (0.998), Pa-234 (0.002)
Te-129m	Te-129 (0.65)	U-240	Np-240m (1.0)
Te-131m	Te-131 (0.222)	Am-242m	Am-242 (0.9952), Np-238 (0.0048)
Te-132	I-132 (1.0)	Am-243	Np-239 (1.0)
Te-133m	Te-133 (0.13)	Cm-247	Pu-243 (1.0)
Te-134	I-134 (1.0)	Es-254	Bk-250 (1.0)

RASCAL also truncates decay chains for deposited radionuclides in intermediate phase dose calculations according to the following rules. Daughters that are noble gases are truncated from the chain because it is assumed that the noble gas will become airborne and be carried away. The exception is when the noble gas is short-lived so that it is included implicitly with the parent nuclide. For example, Xe-135m (half-life of 15 minutes) is included implicitly with its parent I-135, but the chain is truncated at that point because the next nuclide in the chain, Xe-135, is a noble gas with a relatively long half-life (9 hours).

Decay chains are also usually truncated at the first very long-lived daughter in the chain (long relative to the 50-year intermediate phase period) because it is assumed that the ingrowth of the daughter will not cause sufficient concentrations of the daughter to contribute significantly to dose relative to the parent. The exception is that Np-239 decay does include Pu-239 as a member of the chain because of the potential dose importance of decay. Most of the simplification in the RASCAL decay chains occurs for high atomic number parents (Rn and higher).

Table 7.2 lists the decay chains involving more than one explicit radionuclide. Radionuclides having implicit daughters are shown in bold. All other radionuclides are assumed to decay to a stable isotope.

With these simplifications, no intermediate phase decay chain includes more than two generations of explicit daughters. With branching, those two generations of daughters include up to three explicit daughters in a few cases.

Table 7.2 RASCAL Intermediate Phase Decay Chains

Zn-69m[1] –> stable
Ge-68 –> stable
Br-83 –> stable
Rb-89 –> Sr-89 –> stable
Sr-90 –> Y-90 –> stable
Sr-91 –> Y-91 –> stable
Y-91m –> Y-91 –> stable
Y-93 –> Zr-93 –> stable
Zr-93 –> stable
Zr-95 –> Nb-95m –> Nb-95 –> stable
Zr-97 –> stable
Nb-95m –> Nb-95 –> stable
Nb-97m –> Nb-97 –> stable
Mo-93 –> Nb-93m –> stable
Mo-99 –> Tc-99 –> stable
Tc-99m –> Tc-99 –> stable
Ru-103 –> stable
Ru-105 –> Rh-105 –> stable
Ru-106 –> stable
Pd-103 –> stable
Ag-110m –> stable
Cd-115 –> stable
In-114m –> stable
Sn-113 –> stable
Sn-121m –> Sn-121 –> stable
Sn-125 –> Sb-125 –> Te-125m –> stable
Sn-126 –> stable
Sb-125 –> Te-125m –> stable
Sb-126m –> Sb-126 –> stable
Sb-127 –> Te-127m –> Te-127 –> stable
Te-127m –> Te-127 –> stable
Te-129m –> I-129 –> stable
Te-129 –> I-129 –> stable
Te-131m –> I-131 –> stable
Te-131 –> I-131 –> stable
Te-132 –> stable
Te-133m –> I-133 –> stable
Te-133 –> I-133 –> stable
Te-134 –> stable
I-135 –> stable
Cs-134m –> Cs-134 –> stable
Cs-137 –> stable
Ba-140 –> La-140 –> stable
Ba-141 –> La-141 –> Ce-141 –> stable
Ba-142 –> La-142 –> stable
La-141 –> Ce-141 –> stable
Ce-143 –> Pr-143 –> stable
Ce-144 –> stable
Nd-147 –> Pm-147 –> stable
Pm-148m –> Pm-148 –> stable
Pm-151 –> Sm-151 –> stable

W-187 –> Re-187 –> stable
Pb-210 –> Bi-210 –> Po-210 –> stable
Pb-211 –> stable
Pb-212 –> stable
Bi-210 –> Po-210 –> stable
Bi-212 –> stable
Rn-220 –> stable
Rn-222 –> stable
Ra-223 –> stable
Ra-224 –> **Pb-212** –> stable
Ra-225 –> **Ac-225** –> stable
Ra-226 –> stable
Ra-228 –> Th-228 –> **Ra-224** –> stable
Ac-225 –> stable
Ac-227 –> Th-227 –> **Ra-223** –> stable
Th-227 –> **Ra-223** –> stable
Th-228 –> **Ra-224** –> **Pb-212** –> stable
Th-229 –> stable
Th-230 –> **Ra-226** –> stable
Th-231 –> Pa-231–> stable
Th-232 –> **Ra-228** –> Th-228 –> stable
Th-234 –> U-234 –> stable
Pa-233 –> U-233 –> Th-229 –> stable
U-232 –> Th-228 –> **Ra-224** –> stable
U-233 –> Th-229 –> stable
U-234 –> Th-230 –> stable
U-235 –> Pa-231 –> stable
U-237 –> Np-237 –> Pa-233 –> stable
U-238 –> **Th-234** –> U-234 –> stable
U-240 –> stable
Np-237 –> Pa-233 –> U-233 –> stable
Np-239 –> Pu-239 –> stable
Pu-237 –> Np-237 –> stable
Pu-238 –> U-234 –> stable
Pu-241 –> Am-241–> stable
Pu-244 –> Pu-240 –> stable
Am-242m –> Cm-242 –> Pu-238–> stable (0.823)[2]
       –> Pu-242 –> stable (0.172)[2]
Am-242 –> Cm-242 –> Pu-238 –> stable
Am-243 –> Pu-239 –> stable
Cm-242 –> Pu-238 –> stable
Cm-243 –> **Am-243** –> stable (0.0024)[2]
       –> Pu-239 –> stable (0.9976)[2]
Cm-244 –> Pu-240 –> stable
Cm-245 –> Pu-241 –> Am-241 –> stable
Cm-247 –> **Am-243** –> Pu-239 –> stable
Es-254 –> Cf-250 –> Cm-246 –> stable

[1] Bold indicates implicit daughters included as shown in Table 7.1.
[2] Branching fraction

The following equations describe the activity of the parent and explicit daughters as a function of time. Given the measured activity of a radionuclide at time zero, the activities of the radionuclide and daughters, if any, following the measurement are given by the Bateman Equations (e.g., Benedict et al. 1987), modified to give activities (disintegrations/time) rather than number of atoms (Strenge 1997). The Bateman Equation for the parent radionuclide is

$$A_p(t) = A_p(0)e^{-\lambda_p t} \tag{7.13}$$

where

$A_p(t)$ = activity of the parent at time t following the measurement,

$A_p(0)$ = measured activity, and

λ_p = decay constant of the parent.

The activity of first-generation explicit daughter radionuclides, assuming none of the daughter is present initially, is

$$A_{d1}(t) = A_p(0)\frac{f_{d1}\lambda_{d1}}{\lambda_{d1} - \lambda_p}\left(e^{-\lambda_p t} - e^{-\lambda_{d1} t}\right) \tag{7.14}$$

where

$A_{d1}(t)$ = activity of the first-generation daughter at time t,

f_{d1} = fraction of disintegrations of the parent that yield the daughter

λ_{d1} = decay constant for the daughter.

This relationship applies to each first-generation daughter of the parent.

Although there may be more than one second-generation daughter for a parent radionuclide, RASCAL assumes simplified decay chains that include, at most, a single second-generation daughter. The activity of second-generation daughter radionuclides, assuming none of the preceding first generation daughter and none second-generation daughter are present initially, is

$$A_{d2}(t) = A_p(0) \times f_{d1}\lambda_{d1}f_{d2}\lambda_{d2} \times$$
$$\left[\frac{e^{-\lambda_p t}}{(\lambda_{d1} - \lambda_p)(\lambda_{d2} - \lambda_p)} + \frac{e^{-\lambda_{d1} t}}{(\lambda_p - \lambda_{d1})(\lambda_{d2} - \lambda_{d1})} + \frac{e^{-\lambda_{d2} t}}{(\lambda_p - \lambda_{d2})(\lambda_{d1} - \lambda_{d2})}\right] \tag{7.15}$$

where

$A_{d2}(t)$ = activity of the second-generation daughter at time t following the measurement,

7-10

f_{d2} = fraction of first-generation daughter disintegrations that yield the second generation daughter, and

λ_{d2} = decay constant for the second-generation daughter.

In the above equations, the time-dependent activities of the parent and daughters are functions of the initial (measured) activity of the parent.

7.3 Derived Response Levels (DRLs)

A derived response level is a measurable quantity that indicates that the deposited activity could result in an intermediate phase dose equal to one of the intermediate phase protective action guides. One type of DRL is the closed window (gamma) dose rate in mR/hour equal to the first, second, or 50 year intermediate phase PAG. The other DRL is the ground concentration of a marker radionuclide equivalent to a PAG.

RASCAL computes two sets of DRLs - one set for use with measured exposure rates (meter readings), and the other for use with measurement of the surface contamination of a marker radionuclide. In either case, the DRLs are based on an assumed mixture of radionuclides on the surface. In RASCAL, DRLs are computed for first year, second year, and 50 years for a range of times from 0 to 100 days since the initial measurement.

7.3.1 Exposure Rate Derived Response Levels

The exposure rate derived response level (DRL) is the exposure rate that will occur when the ground concentration present will cause a dose to an inhabitant that is equal to the intermediate phase protective action guide. Thus, an exposure rate measurement survey instrument can be used to identify areas where doses might exceed the intermediate phase protective action guides.

Exposure rate DRLs are calculated using equations developed from methods described in the FRMAC Manual (SNL 2003). However, the RASCAL DRLs include an inadvertent ingestion component that is not included in the FRMAC computational method. The basic equation used in RASCAL for exposure rate DRLs is

$$DRL_{exp} = PAG \times \frac{1.429 \times \sum_{i}^{n} \left(C_{gi} \times GRF \times ECF_{egi} \times 3600. \right)}{\sum_{i}^{n} C_{gi} \left[\left(sfc\_rf \times DCP_{esi} \right) + \left(ING \times DCP_{ingi} \right) + \left(R_{so} \times V_{b} \times DCP_{inhi} \right) \right]} \qquad (7.16)$$

where

DRL_{exp} = derived response level for exposure. (mR/hr)

PAG = EPA protective action guide. (mrem)

1.429 = conversion factor from mrem to mR from EPA 1992, page 7-11

ECF_{egi} = exposure rate dose coefficient for contaminated soil from Table III.3 of FGR No. 12 (Eckerman and Ryman 1993)

3600 s/hr converts the exposure rate to an hourly rate.

The occupancy time interval for the intermediate phase dose conversion parameters are usually for the occupancy time intervals established for the EPA protective action guides for the intermediate phase.

7.3.2 Marker Nuclide Derived Response Levels

In some instances it may be difficult to use the exposure rate DRL to identify areas where doses might exceed the protective action guides. Examples are when the exposure rate is near background levels or when there are no gamma-emitting radionuclides in the mix. In those instances, it may be easier to measure the surface concentration of a particular marker radionuclide rather than the exposure rate.

The marker nuclide DRL is the concentration of that nuclide expected to cause doses to inhabitants equal to the protective action guides. Following the method in the FRMAC Manual (SNL 2003), RASCAL calculates marker DRLs as follows

$$DRL_{mark} = PAG \times \frac{C_{gk}}{\sum_i^n C_{gi}\left[\left(GRF \times DCP_{gsi}\right) + \left(ING \times DCP_{ingi}\right) + \left(R_{so} \times V_b \times DCP_{inhi}\right)\right]} \qquad (7.17)$$

where

C_{gk} = ground concentration of the marker radionuclide (mCi/m^2)

As with the exposure rate DRLs, the RASCAL calculation includes an optional inadvertent ingestion component not included in the FRMAC DRL calculation.

7.4 References

Anspaugh, L . R., S. L. Simon, K. I. Gordeev, I. A. Likhtarev, R. M. Maxwell, and S. M. Shinkarev. 2002. "Movement of Radionuclides in Terrestrial Ecosystems by Physical Processes. *Health Physics* 82(5):669-679.

Benedict, M., T. H. Pigford, and H. W. Levi. 1987. *Nuclear Chemical Engineering*, 2nd Ed. McGraw-Hill, New York.

Eckerman, K.F., A. B. Wobarst, and A.C. B.Richardson. 1988. *Limiting Values of Radionuclide Intake and Air Concentration and Dose Conversion Factors for Inhalation, Submersion, and Ingestion.* Federal Guidance Report No. 11. EPA-520/11-88-020, U.S. Environmental Protection Agency.

Eckerman, K. F. and J. C. Ryman. *1993. External Exposure to Radionuclides in Air Water, and Soil.* Federal Guidance Report No. 12. EPA-402-R-93-081, U.S. Environmental Protection Agency.

National Council on Radiation Protection and Measurements (NCRP). 1999. *Recommended Screening Limits for Contaminated Surface Soil and Review of Factors Relevant to Site-Specific Studies.* NRCP Report No. 129. National Council on Radiation Protection and Measurements, Bethesda, Maryland.

Sandia National Laboratories (SNL). 2003. *FRMAC Assessment Manual, Volume 1 Methods.* SAND2003-107P, Sandia National Laboratories, Albuquerque, New Mexico.

Strenge, D. L. 1997. "A General Algorithm for Radioactive Decay With Branching and Loss from a Medium," *Health Physics* 73:953-957.

U.S. Environmental Protection Agency. 1992. *Manual of Protective Action Guides and Protective Actions for Nuclear Incidents.* EPA 400-R-92-001.

NRC FORM 335
(9-2004)
NRCMD 3.7

U.S. NUCLEAR REGULATORY COMMISSION

BIBLIOGRAPHIC DATA SHEET

(See instructions on the reverse)

1. REPORT NUMBER (Assigned by NRC, Add Vol., Supp., Rev., and Addendum Numbers, if any.)
NUREG-1887

2. TITLE AND SUBTITLE	3. DATE REPORT PUBLISHED	
RASCAL 3.0.5: Description of Models and Methods	MONTH	YEAR
	August	2007
	4. FIN OR GRANT NUMBER	
	R1110	

5. AUTHOR(S)	6. TYPE OF REPORT
Stephen A. McGuire James Van Ramsdell, Pacific Northwest Laboratories George F. Athey, Athey Consulting	Final technical report
	7. PERIOD COVERED *(Inclusive Dates)*
	NA

8. PERFORMING ORGANIZATION - NAME AND ADDRESS *(If NRC, provide Division, Office or Region, U.S. Nuclear Regulatory Commission, and mailing address; if contractor, provide name and mailing address.)*

Divsion of Preparedness and Response
Office of Nuclear Security and Incident Response
U. S. Nuclear Regulatory Commission
Washington, DC 20555

9. SPONSORING ORGANIZATION - NAME AND ADDRESS *(If NRC, type "Same as above"; if contractor, provide NRC Division, Office or Region, U.S. Nuclear Regulatory Commission, and mailing address.)*

same as above

10. SUPPLEMENTARY NOTES
Supercedes NUREG-1741.

11. ABSTRACT *(200 words or less)*

The code currently used by NRC's emergency operations center for making dose projections for radiological emergencies is RASCAL version 3.0.5 (Radiological Assessment System for Consequence AnaLysis). This code was developed by NRC. The first version was created about 20 years ago. Since then the code has been undergoing continual improvement to expand its capabilities and to update the models used in its calculations. This report describes the models and calculational methods used in RASCAL 3.0.5. This report updates and supercedes the information in NUREG-1741, "RASCAL 3.0: Description of Models and Methods," 2001.

RASCAL 3.0.5 evaluates releases from: nuclear power plants, spent fuel storage pools and casks, fuel cycle facilities, and radioactive material handling facilities.

12. KEY WORDS/DESCRIPTORS *(List words or phrases that will assist researchers in locating the report.)*

RASCAL, emergency response, dose assessment, dose projections, plume modeling

13. AVAILABILITY STATEMENT
unlimited
14. SECURITY CLASS FICATION
(This Page)
unclassified
(This Report)
unclassified
15. NUMBER OF PAGES
16. PRICE

www.ingramcontent.com/pod-product-compliance
Lightning Source LLC
Chambersburg PA
CBHW081454170526
45166CB00008B/2422